BARRON'S

Regents Exams and Answers

Algebra I

Revised Edition

GARY M. RUBINSTEIN
B.A. Mathematics
Tufts University
M.S. Computer Science
University of Colorado

D0910684

Kaplan North America, LLC, dba Barron's Educational Series
1515 West Cypress Creek Road
Fort Lauderdale, Florida 33309
www.barronseduc.com

ISBN: 978-1-5062-6633-6

10 9 8 7 6 5 4 3

Kaplan North America, LLC, dba Barron's Educational Series print books are available at
special quantity discounts to use for sales promotions, employee premiums, or educational
purposes. For more information or to purchase books, please call the Simon & Schuster
special sales department at 866-506-1949.

Contents

Regents Examinations, Answers, and Self-Analysis Charts   165

Preface

In June of 2014, New York State adopted a more difficult Algebra I course and changed the Regents exam to reflect this new course. On the Algebra I Regents exam, students are expected not only to solve algebra problems and answer algebra-related questions but also to *explain their reasoning*. This is a major shift in the testing of algebra concepts, so a thorough review is even more critical for success.

This book will introduce you to the concepts covered throughout the Algebra I Regents course as well as to the different types of problems you will encounter on the Regents exam at the end of the year.

WHO SHOULD USE THIS BOOK?

Students and teachers alike can use this book as a resource for preparing for the Algebra Regents exam.

Students will find this book to be a great study tool because it contains a review of Algebra I concepts, useful examples, and practice problems of varying difficulty that can be practiced throughout the school year to reinforce what they are learning in class. The most ideal way to prepare for the Algebra Regents exam is to work through the practice problems in the review sections and then the recently administered Regents exams at the end of the book.

Teachers can use this book as a tool to help structure an Algebra I course that will culminate with the Regents exam. The topics in the book are arranged by priority, so the sections in the beginning of the book are the ones from which more of the questions on the test are drawn. There are 13 sections, dedicated to all topics for Algebra I, each with practice exercises and solutions.

WHY IS THIS BOOK A HELPFUL RESOURCE?

Becoming familiar with the specific types of questions on the Algebra I Regents exam is crucial to performing well on this test. There are questions in which the math may be fairly easy but the way in which the question is asked makes the question seem much more difficult. For example, the question "Find all zeros of the function $f(x) = 2x + 6$" is a fancy way of asking the much simpler sounding "Solve for x if $2x + 6 = 0$." *Knowing exactly what the questions are asking is a big part of success on this test.*

Algebra has been around for thousands of years, and its basic concepts have never changed. So fundamentally, this algebra curriculum is not very different from the algebra taught in schools two years ago, ten years ago, or twenty years ago. But the exam, with more complicated ways of asking questions and presenting problems, requires a specifically presented study plan that's more important than ever.

Gary Rubinstein

How to Use This Book

This book is designed to help you get the most out of your review for the new Regents exam in Algebra I. Use this book to improve your understanding of the Algebra I topics and improve your grade.

TEST-TAKING TIPS

The first section in this book contains test-taking tips and strategies to help prepare you for the Algebra I Regents exam. This information is valuable, so be sure to read it carefully and refer to it during your study time. Remember: no single problem-solving strategy works for all problems—you should have a toolbox of strategies to pick from as you're facing unfamiliar or difficult problems on the test.

PRACTICE WITH KEY ALGEBRA I FACTS AND SKILLS

The second section in this book provides you with key Algebra I facts, useful skills, and practice problems with solutions. It provides you with a quick and easy way to refresh the skills you learned in class.

REGENTS EXAMS AND ANSWERS

The final section of the book contains actual Algebra I Regents exams that were administered in June 2017, August 2017, June 2018, August 2018, June 2019, and August 2019. These exams and thorough answer explanations are probably the most useful tool for your review, as they let you know what's most important. By answering the questions on these exams, you will be able to identify your strengths and weaknesses and then concentrate on the areas in which you may need more study.

Remember, the answer explanations in this book are more than just simple solutions to the problems—they contain facts and explanations that are crucial to success in the Algebra I course and on the Regents exam. Careful review of these answers will increase your chances of doing well.

SELF-ANALYSIS CHARTS

Each of the Algebra I Regents exams ends with a Self-Analysis Chart. This chart will further help you identify weaknesses and direct your study efforts where needed. In addition, the chart classifies the questions on each exam into an organized set of topic groups. This feature will also help you to locate other questions on the same topic in the other Algebra I exams.

IMPORTANT TERMS TO KNOW

The terms that are listed in the glossary are the ones that have appeared most frequently on past Integrated Algebra and the most recently Algebra I exams. All terms and their definitions are conveniently organized for a quick reference.

Test-Taking Tips and Strategies

Knowing the material is only part of the battle in acing the new Algebra I Regents exam. Things like improper management of time, careless errors, and struggling with the calculator can cost valuable points. This section contains some test-taking strategies to help you perform your best on test day.

Tip 1
Time Management

SUGGESTIONS

- *Don't rush.* The Algebra I Regents exam is three hours long. While you are officially allowed to leave after 90 minutes, you really should stay until the end of the exam. Just as it wouldn't be wise to come to the test an hour late, it is almost as bad to leave a test an hour early.

- *Do the test twice.* The best way to protect against careless errors is to do the entire test twice and compare the answers you got the first time to the answers you got the second time. For any answers that don't agree, do a "tie breaker" third attempt. Redoing the test and comparing answers is much more effective than simply looking over your work. Students tend to miss careless errors when looking over their work. By redoing the questions, you are less likely to make the same mistake.

- *Bring a watch.* What will happen if the clock is broken? Without knowing how much time is left, you might rush and make careless errors. Yes, the proctor will probably write the time elapsed on the board and update it every so often, but its better safe than sorry.

The TI-84 graphing calculator has a built-in clock. Press the [MODE] to see it. If the time is not right, go to SET CLOCK and set it correctly. The TI-Nspire does not have a built-in clock.

 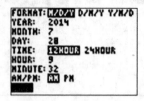

TIP 2

Know How to Get Partial Credit

SUGGESTIONS

- *Know the structure of the exam.* The Algebra Regents exam has 37 questions. The first 24 of those questions are multiple-choice worth two points each. There is no partial credit if you make a mistake on one of those questions. Even the smallest careless error, like missing a negative sign, will result in no credit for that question. Parts Two, Three, and Four are free-response questions with no multiple-choice. Besides giving a numerical answer, you may be asked to explain your reasoning. Part Two has eight free-response questions worth two points each. The smallest careless error will cause you to lose one point, which is half the value of the question. Part Three has four free-response questions worth four points each. These questions generally have multiple parts. Part Four has one free-response question worth six points. This question will have multiple parts.

- *Explain your reasoning.* When a free-response question asks to "Justify your answer," "Explain your answer," or "Explain how you determined your answer," the grader is expecting a few clearly written sentences. For these, you don't want to write too little since the grader needs to see that you understand why you did the different steps you did to solve the equation. You also don't want to write too much because if anything you write is not accurate, points can be deducted.

Here is an example followed by two solutions. The first would not get full credit, but the second would.

Example 1

Use algebra to solve for x in the equation $\frac{2}{3}x + 1 = 11$. Justify your steps.

Solution 1 (part credit):

$\frac{2}{3}x + 1 = 11$ $-1 = -1$ $\frac{2}{3}x = 10$ $x = 15$	I used algebra to get the x by itself. The answer was $x = 15$.

Solution 2 (full credit):

$\frac{2}{3}x + 1 = 11$ $-1 = -1$ $\frac{2}{3}x = 10$ $\frac{3}{2} \cdot \frac{2}{3}x = \frac{3}{2} \cdot 10$ $1x = 15$ $x = 15$	I used the subtraction property of equality to eliminate the $+1$ from the left-hand side. Then to make it so the x had a 1 in front of it, I used the multiplication property of equality and multiplied both sides of the equation by the reciprocal of $\frac{2}{3}$, which is $\frac{3}{2}$. Then since $1 \cdot x = x$, the left-hand side of the equation just became x and the right-hand side became 15.

- *Computational errors vs. conceptual errors*

  In the Part Three and Part Four questions, the graders are instructed to take off one point for a "computational error" but half credit for a "conceptual error." This is the difference between these two types of errors.

  If a four-point question was $x - 1 = 2$ and a student did it like this,

$$x - 1 = 2$$
$$+1 = +1$$
$$x = 4$$

the student would lose one point out of 4 because there was one computational error since $2 + 1 = 3$ and not 4.

Had the student done it like this,

$$x - 1 = 2$$
$$-1 = -1$$
$$x = 1$$

the student would lose half credit, or 2 points, since this error was conceptual. The student thought that to eliminate the -1, he should subtract 1 from both sides of the equation.

Either error might just be careless, but the conceptual error is the one that gets the harsher deduction.

TIP 3

Know Your Calculator

SUGGESTIONS

- *Which calculator should you use?* The two calculators used for this book are the TI-84 and the TI-Nspire. Both are very powerful. The TI-84 is somewhat easier to use for the functions needed for this test. The TI-Nspire has more features for courses in the future. The choice is up to you. This author prefers the TI-84 for the Algebra Regents. Graphing calculators come with manuals that are as thick as the book you are holding. There are also plenty of video tutorials online for learning how to use advanced features of the calculator. To become an expert user, watch the online tutorials or read the manual.

- *Clearing the memory.* You may be asked at the beginning of the test to clear the memory of your calculator. When practicing for the test, you should clear the memory too so you are practicing under test-taking conditions.

  This is how you clear the memory.

For the TI-84:

Press [2ND] and then [+] to get to the MEMORY menu. Then press [7] for Reset.

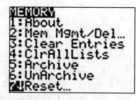

Use the arrows to go to [ALL] for All Memory. Then press [1].

Press [2] for Reset.

The calculator will be reset as if in brand new condition. The one setting that you may need to change is to turn the diagnostics on if you need to calculate the correlation coefficient.

For the TI-Nspire:

The TI-Nspire must be set to Press-To-Test mode when taking the Algebra Regents. Turn the calculator off by pressing [ctrl] and [home]. Press and hold [esc] and then press [home].

While in Press-to-Test mode, certain features will be deactivated. A small green light will blink on the calculator so a proctor can verify the calculator is in Press-to-Test mode.

To exit Press-to-Test mode, use a USB cable to connect the calculator to another TI-Nspire. Then from the home screen on the calculator in Press-to-Test mode, press [doc], [9] and select Exit Press-to-Test.

- *Use parentheses*
  The calculator always uses the order of operations where multiplication and division happen before addition and subtraction. Sometimes, though, you may want the calculator to do the operations in a different order.

  Suppose at the end of a quadratic equation, you have to round $x = \dfrac{-1 + \sqrt{5}}{2}$ to the nearest hundredth. If you enter $(-)$ (1) (+) (2ND) (x^2) (5) (/) (2), it displays

 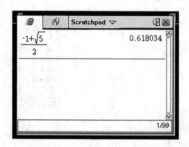

  which is not the correct answer.

  One reason is that for the TI-84 there needs to be a closing parenthesis (or on the TI-Nspire, press [right arrow] to move out from under the radical sign) after the 5 in the square root symbol. Without it, it calculated $-1 + \sqrt{\dfrac{5}{2}}$. More needs to be done, though, since

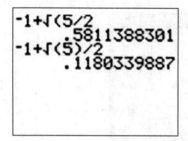

  still is not correct. This is the solution to $-1 + \sqrt{\dfrac{5}{2}}$.

  To get this correct, there also needs to be parentheses around the entire numerator, $-1 + \sqrt{5}$.

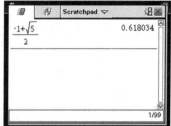

This is the correct answer.

On the TI-Nspire, fractions like this can also be done with [templates].

 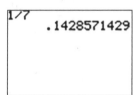

- *Using the ANS feature*

The last number calculated with the calculator is stored in something called the ANS variable. This ANS variable will appear if you start an expression with a $+$, $-$, $\times$, or $\div$. When an answer has a lot of digits in it, this saves time and is also more accurate.

If for some step in a problem you need to calculate the decimal equivalent of $\frac{1}{7}$, it will look like this on the TI-84:

```
1/7
     .1428571429
```

For the TI-Nspire, if you try the same thing, it leaves the answer as $\frac{1}{7}$. To get the decimal approximation, press [ctrl] and [enter] instead of just [enter].

Now if you want to multiply this by 3, just press [×], and the calculator will display "Ans*"; press [3] and [enter].

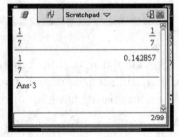

The ANS variable can also help you do calculations in stages. To calculate $x = \dfrac{-1 + \sqrt{5}}{2}$ without using so many parentheses as before, it can be done by first calculating $-1 + \sqrt{5}$ and then pressing [÷] and [2] and Ans will appear automatically.

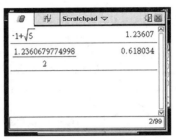

The ANS variable can also be accessed by pressing [2ND] and [−] at the bottom right of the calculator. If after calculating the decimal equivalent of $\frac{1}{7}$ you wanted to subtract $\frac{1}{7}$ from 5, for the TI-84 press [5], [−], [2ND], [ANS], and [ENTER]. For the TI-Nspire press [5], [−], [ctrl], [ans], and [enter].

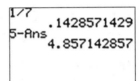

TIP 4

Use the Reference Sheet

SUGGESTIONS

- In the back of the Algebra Regents booklet is a reference sheet that contains 17 conversion facts, such as inches to centimeters and quarts to pints, and also 17 formulas. Many of these conversion facts and formulas will not be needed for an individual test, but the quadratic formula and the arithmetic sequence formula are the two that will come in the handiest.

High School Math Reference Sheet

1 inch = 2.54 centimeters	1 kilometer = 0.62 mile	1 cup = 8 fluid ounces
1 meter = 39.37 inches	1 pound = 16 ounces	1 pint = 2 cups
1 mile = 5280 feet	1 pound = 0.454 kilogram	1 quart = 2 pints
1 mile = 1760 yards	1 kilogram = 2.2 pounds	1 gallon = 4 quarts
1 mile = 1.609 kilometers	1 ton = 2000 pounds	1 gallon = 3.785 liters
		1 liter = 0.264 gallon
		1 liter = 1000 cubic centimeters

Triangle	$A = \frac{1}{2}bh$	Pythagorean Theorem	$a^2 + b^2 = c^2$
Parallelogram	$A = bh$	Quadratic Formula	$x = \dfrac{-b \pm \sqrt{b^2 - 4ac}}{2a}$
Circle	$A = \pi r^2$	Arithmetic Sequence	$a_n = a_1 + (n-1)d$
Circle	$C = \pi d$ or $C = 2\pi r$	Geometric Sequence	$a_n = a_1 r^{n-1}$
General Prisms	$V = Bh$	Geometric Series	$S_n = \dfrac{a_1 - a_1 r^n}{1 - r}$ where $r \neq 1$
Cylinder	$V = \pi r^2 h$	Radians	$1 \text{ radian} = \dfrac{180}{\pi} \text{ degrees}$
Sphere	$V = \frac{4}{3}\pi r^3$	Degrees	$1 \text{ degree} = \dfrac{\pi}{180} \text{ radians}$
Cone	$V = \frac{1}{3}\pi r^2 h$	Exponential Growth/Decay	$A = A_0 e^{k(t - t_0)} + B_0$
Pyramid	$V = \frac{1}{3}Bh$		

- *How Many Points Do You Need to Pass?*

The Algebra Regents exam is scored out of a possible 86 points. Unlike most tests given in the year by your teacher, the score is not then turned into a percent out of 86. Instead each test has a conversion sheet that varies from year to year. For example, for the June 2014 test, the conversion sheet looked like this.

Raw Score	Scale Score	Raw Score	Scale Score	Raw Score	Scale Score
86	100	57	75	28	64
85	99	56	74	27	63
84	97	55	74	26	62
83	96	54	74	25	61
82	95	53	73	24	60
81	94	52	73	23	59
80	92	51	73	22	58
79	91	50	72	21	56
78	90	49	72	20	55
77	89	48	72	19	54
76	88	47	72	18	52
75	87	46	71	17	50
74	86	45	71	16	49
73	85	44	71	15	47
72	84	43	70	14	45
71	83	42	70	13	42
70	82	41	70	12	40
69	82	40	70	11	38
68	81	39	69	10	35
67	80	38	69	9	32
66	79	37	69	8	30
65	79	36	68	7	26
64	78	35	68	6	23
63	78	34	67	5	20
62	77	33	67	4	16
61	77	32	66	3	12
60	76	31	66	2	9
59	76	30	65	1	4
58	75	29	64	0	0

On this test, 30 points became a 65, 57 points became a 75, and 73 points became an 85. This means that for this examination a student who got 30 out of 86, which is just 35% of the possible points, would get a 65 on this exam. 57 out of 86 is 66%, but this scaled to a 75. 73 out of 86, however, is actually 85% and became an 85. So in the past there has been a curve on the exam for lower scores, though the scaling is not released until after the exam.

Some Key Algebra I Facts and Skills

1. PROPERTIES OF ALGEBRA AND SOLVING LINEAR EQUATIONS WITH ALGEBRA

1.1 ONE-STEP ALGEBRA PROBLEMS

An algebra problem, like $x - 5 = 2$, is one that can be solved by changing both sides of the equation until the variable x is isolated. There are four main properties that can be used in solving algebra problems.

- The **addition property** of equality

    The equation $x - 5 = 2$ is solved by adding 5 to both sides of the equation. When you add to both sides of an equation, you are using the addition property of equality.

$$x - 5 = 2 \quad \text{the given equation}$$
$$+5 = +5 \quad \text{addition property of equality}$$
$$x = 7 \quad x \text{ is isolated. The solution is 7.}$$

- The **subtraction property** of equality

    The equation $x + 2 = 7$ is solved by subtracting 2 from both sides of the equation. When you subtract from both sides of an equation, you are using the subtraction property of equality.

$$x + 2 = 7 \quad \text{the given equation}$$
$$-2 = -2 \quad \text{subtraction property of equality}$$
$$x = 5 \quad x \text{ is isolated. The solution is 5.}$$

- The **division** property of equality

  The equation $2x = 10$ is solved by dividing both sides of the equation by 2. When you divide both sides of the equation, you are using the division property of equality.

  $$2x = 10 \quad \text{the given equation}$$

  $$\frac{2x}{2} = \frac{10}{2} \quad \text{division property of equality}$$

  $$x = 5 \quad \text{x is isolated. The solution is 5.}$$

- The **multiplication** property of equality

  The equations $\frac{x}{5} = 3$ and $\frac{2}{3}x = 8$ can be solved by multiplying both sides of the equation by the same number.

  $$\frac{x}{5} = 3 \quad \text{the given equation}$$

  $$5 \cdot \frac{x}{5} = 5 \cdot 3 \quad \text{multiplication property of equality}$$

  $$x = 15 \quad \text{x is isolated. The solution is 5.}$$

  $$\frac{2}{3}x = 8 \quad \text{the given equation}$$

  $$\frac{3}{2} \cdot \frac{2}{3}x = \frac{3}{2} \cdot 8 \quad \text{multiplication property of equality}$$

  $$x = 12 \quad \text{x is isolated. The solution is 12.}$$

1.2 TWO-STEP ALGEBRA PROBLEMS

When an equation has the form $mx + b = y$, it takes two steps to solve for x. The first step is to eliminate the b, which is called the **constant**. The second step is to eliminate the m, which is called the **coefficient**. The b is eliminated with either the addition or the subtraction property of equality. The m is eliminated with either the division or the multiplication property of equality.

- For the equation $3x - 7 = 11$

$3x - 7 = 11$	the given equation
$+7 = +7$	addition property of equality
$\dfrac{3x}{3} = \dfrac{18}{3}$	division property of equality
$x = 6$	x is isolated. The solution is 6.
$\dfrac{2}{3}x + 5 = 11$	the given equation
$-5 = -5$	subtraction property of equality
$\dfrac{2}{3}x = 6$	
$\dfrac{3}{2} \cdot \dfrac{2}{3}x = \dfrac{3}{2} \cdot 6$	multiplication property of equality
$x = 9$	x is isolated. The solution is 9.

1.3 COMBINING LIKE TERMS BEFORE SOLVING

If the equation has multiple x terms or multiple constants, the equation should first be simplified by combining like terms. After all like terms have been combined, the question will usually be a two-step algebra problem and can be solved with the methods from Section 1.2.

$2x + 5 + 3x - 2 = 23$	the given equation
$2x + 3x + 5 - 2 = 23$	Terms are rearranged so the x terms are together and the constants are together. This step is optional.
$5x + 3 = 23$	Like terms have been combined.
$-3 = -3$	subtraction property of equality
$\dfrac{5x}{5} = \dfrac{20}{5}$	division property of equality
$x = 4$	x is isolated. The solution is 4.

1.4 VARIABLES ON BOTH SIDES OF THE EQUATION

When there are x terms on both sides of the equation, the addition property of equality or the subtraction property of equality can be used to change the equation into one where the x terms are all on the same side of the equation.

$$5x - 3 = 12 + 2x \quad \text{the given equation}$$

$$-2x = -2x \qquad \text{subtraction property of equality eliminates the } x \text{ term from the right-hand side of the equation}$$

$$3x - 3 = 12$$

$$+3 = +3 \qquad \text{addition property of equality}$$

$$\frac{3x}{3} = \frac{15}{3} \qquad \text{division property of equality}$$

$$x = 5 \qquad x \text{ is isolated. The solution is 5.}$$

1.5 EQUATIONS WITH MORE THAN ONE VARIABLE

An equation with more than one variable can be solved the same way as an equation with just one variable. The **solution** will not be a number in these problems, but an **expression** with variables and numbers in it.

$$ax + 2 = c \qquad \text{the given equation}$$

$$-2 = -2 \qquad \text{subtraction property of equality}$$

$$ax = c - 2 \qquad \text{the } c \text{ and the 2 cannot be combined since they are unlike terms}$$

$$\frac{ax}{a} = \frac{c - 2}{a} \qquad \text{division property of equality}$$

$$x = \frac{c - 2}{a} \qquad x \text{ is isolated. The solution for } x \text{ is not a number but an expression in terms}$$

of c and a. The answer is $\dfrac{c - 2}{a}$.

Practice Exercises

1. Antonio started the question $2x + 1 = 11$ by writing $2x = 10$. Which property justifies this step?

   (1) Commutative property of addition

   (2) Distributive property of multiplication over addition

   (3) Addition property of equality

   (4) Subtraction property of equality

2. Mila used the multiplication property to justify the first step in solving an equation. The original equation was $\frac{x}{2} + 4 = 10$. What could the equation have been transformed into after this step?

   (1) $\frac{x}{2} = 6$ (3) $x + 8 = 20$

   (2) $\frac{x}{2} + 6 + 12$ (4) $x + 2 = 5$

3. What is the solution set for the equation $x - 6 = 7$?

   (1) $\{1\}$ (3) $\{11\}$

   (2) $\{6\}$ (4) $\{13\}$

4. What value of x makes the equation $3x + 7 = 22$ true?

   (1) 1   (3) 5

   (2) 3   (4) 7

5. Find the solution set for the equation $5(x + 4) = 35$.

   (1) $\{1\}$ (3) $\{3\}$

   (2) $\{2\}$ (4) $\{4\}$

6. Solve for d in terms of c, e, and f.
$$cd - e = f$$

(1) $\dfrac{f - e}{c}$ (3) $\dfrac{f}{c} + e$

(2) $\dfrac{f + e}{c}$ (4) $\dfrac{f}{c} - e$

7. Solve for m in terms of a, b, and c.
$$b - ma = c$$

(1) $\dfrac{b - c}{-a}$ (3) $\dfrac{c - b}{-a}$

(2) $(c - b) - a$ (4) $(c + b) - a$

8. Solve for r in terms of c and π.
$$c = 2\pi r$$

(1) $\dfrac{2c}{\pi}$ (3) $\dfrac{c}{2\pi}$

(2) $\dfrac{2\pi}{c}$ (4) $\dfrac{2}{c\pi}$

Solutions

1. The first step of the process is to subtract 1 from each side of the equation. This is called the subtraction property of equality.

    The correct choice is **(4)**.

2. Though it is more common to begin this question by subtracting 4 from both sides of the equation, in this case she does it by multiplying both sides of the equation by 2. The left-hand side becomes $x + 8$ and the right-hand side becomes 20.

    The correct choice is **(3)**.

3. Isolate the x by adding 6 to both sides of the equation. The equation then becomes $x = 13$.

    The correct choice is **(4)**.

4. Subtract 7 from both sides of the equation to get $3x = 15$. Divide both sides of the equation by 3 to get $x = 5$.

    The correct choice is **(3)**.

5. One way to solve this equation is to first distribute the 5 through the left-hand side to get the equation $5x + 20 = 35$, then subtract 20 from both sides of the equation to get $5x = 15$, and finally to divide both sides of the equation by 5 to get $x = 3$. Another way is to first divide both sides of the equation by 5 to get $x + 4 = 7$, and then subtract 4 from both sides of the equation to get $x = 3$.

    The correct choice is **(3)**.

6. First add e to both sides of the equation to get $cd = f + e$. Then divide both sides of the equation by c to get $d = (f + e)/c$.

   The correct choice is **(2)**.

7. First subtract b from both sides of the equation to get $-ma = c - b$. Then divide both sides by $-a$ to get $m = \dfrac{c - b}{-a}$.

   The correct choice is **(3)**.

8. Divide both sides of the equation by 2π to get $\dfrac{c}{2\pi} = r$.

   The correct choice is **(3)**.

2. POLYNOMIAL ARITHMETIC

2.1 CLASSIFYING POLYNOMIALS

A **polynomial** is an expression like $2x + 5$ or $3x^2 - 5x + 3$. The **terms** of a polynomial are separated by $+$ or $-$ signs. The polynomial $2x + 5$ has two terms. The polynomial $3x^2 - 5x + 3$ has three terms. The terms of a polynomial have a **coefficient** and a **variable part**. The term $3x^2$ has a coefficient of 3 and a variable part of x^2. A term with no variable part is called a **constant**.

- A polynomial with three terms is called a **trinomial**.
- A polynomial with two terms is called a **binomial**.
- A polynomial with one term is called a **monomial**.

2.2 MULTIPLYING AND DIVIDING MONOMIALS

- To *multiply* one monomial by another, multiply the coefficients and multiply the variable parts by adding the exponents on the same variables.

$$8x^3 \cdot 2x$$

Multiply the coefficients $8 \cdot 2 = 16$.

Multiply the variable parts by adding the exponents $x^3 \cdot x^1 = x^4$. The solution is $16x^4$.

- To *divide* one monomial by another, divide the coefficients and divide the variable parts by subtracting the exponents on the same variables.

$$8x^3 \div 2x$$

Divide the coefficients. $8 \div 2 = 4$.

Divide the variable parts by subtracting the exponents

$$x^3 \div x^1 = x^2$$

The solution is $4x^2$.

This question can also be expressed as $\dfrac{8x^3}{2x} = 4x^2$.

2.3 COMBINING LIKE TERMS

Like terms are terms that have the same variable part. For example, $3x^2$ and $2x^2$ are like terms because the variable part for both is x^2. Like terms can be added or subtracted by adding or subtracting the coefficients and by not changing the variable part.

$$3x^2 + 2x^2 = 5x^2$$
$$3x^2 - x^2 = 3x^2 - 1x^2 = 2x^2$$

- To simplify $2x + 3 + 4x - 5$, combine the like terms with the variable part of x. $2x + 4x = 6x$. Also combine the constants $3 - 5 = -2$. This expression simplifies to $6x - 2$.

If terms are not like terms, they cannot be combined by adding or subtracting. $3x^2 + 5x^3$ cannot be combined because the exponents are different, so they are not like terms.

2.4 MULTIPLYING MONOMIALS AND POLYNOMIALS

- To *multiply* a polynomial by a monomial, use the **distributive property**.

$$2(3x + 5) = 2 \cdot 3x + 2 \cdot 5 = 6x + 10$$

This works for more complex monomials and polynomials also.

$$2x^2(5x^2 - 7x + 3) = 2x^2 \cdot 5x^2 + 2x^2 \cdot -7x + 2x^2 \cdot 3$$
$$= 10x^4 - 14x^3 + 6x^2$$

2.5 ADDING AND SUBTRACTING POLYNOMIALS

- To *add* two polynomials, remove the parentheses from both and combine like terms.

$$(5x + 2) + (3x - 4) = 5x + 2 + 3x - 4 = 8x - 2$$

- To **subtract** two polynomials, remove the parentheses of the polynomial on the left, then negate all the terms of the polynomial on the right, and remove the parentheses before combining like terms.

$$(5x + 2) - (3x - 4) = 5x + 2 - 3x + 4 = 2x + 6$$

2.6 MULTIPLYING BINOMIALS

- To *multiply* binomials, use the <u>*FOIL*</u> process.

$$(2x + 3)(5x - 2)$$

The F stands for firsts. Multiply $2x \cdot 5x$, the first term in each of the parentheses. $10x^2$.

The O stands for outers. Multiply $2x \cdot -2$, the terms on the far left and on the far right. $-4x$.

The I stands for inners. Multiply $3 \cdot 5x$, the two terms in the middle. $15x$.

The L stands for lasts. Multiply $3 \cdot -2$, the second term in each of the parentheses. -6

These four answers become $10x^2 - 4x + 15x - 6$. Combine like terms to get $10x^2 + 11x - 6$ in simplified form.

2.7 FACTORING POLYNOMIALS

Factoring a number, like 15, is when two numbers are found that can be multiplied to become that number, $15 = 3 \cdot 5$. Factoring polynomials is more involved and there are certain patterns to be aware of.

- **Greatest Common Factor Factoring**

The terms of some polynomials have a greatest common factor. This can be factored out like a reverse use of the distributive property.

In $6x^2 + 8x$, the terms have a common factor of $2x$. Write $2x$ outside the parentheses and divide each term by $2x$ to determine what goes inside the parentheses.

$$2x(3x + 4)$$

- **Difference of Perfect Squares Factoring**

The expression $a^2 - b^2$ can be factored into $(a - b)(a + b)$. This works anytime both terms of a binomial are perfect squares and there is a minus sign between the two terms.

$$x^2 - 9 = x^2 - 3^2 = (x - 3)(x + 3)$$

- **Reverse FOIL**

A **trinomial** like $x^2 + 8x + 15$ can be factored if there are two numbers that have a sum equal to the coefficient of the x term, 8, and a product equal to the constant 15. Since $3 + 5 = 8$ and $3 \cdot 5 = 15$,

$$x^2 + 8x + 15 = (x + 3)(x + 5).$$

For $x^2 + 3x - 10$, the numbers that have a sum of 3 and a product of -10 are -2 and 5.

$$x^2 + 3x - 10 = (x - 2)(x + 5)$$

2.8 MORE COMPLICATED FACTORING

Sometimes none of the factoring patterns seems to match the polynomial that needs to be factored. When this happens, see if it is possible to rewrite it in a way that resembles the pattern better.

- The polynomial $x^4 - 9$ can be rewritten as $(x^2)^2 - 3^2$, which now has the difference of perfect squares pattern.

$$x^4 - 9 = \left(x^2\right)^2 - 3^2 = \left(x^2 - 3\right)\left(x^2 + 3\right)$$

- The polynomial $x^4 + 8x^2 + 15$ can be rewritten as

$$\left(x^2\right)^2 + 8\left(x^2\right) + 15$$

$$x^4 + 8x^2 + 15 = \left(x^2\right)^2 + 8\left(x^2\right) + 15 = \left(x^2 + 3\right)\left(x^2 + 5\right)$$

Practice Exercises

1. Classify this polynomial $5x^2 + 3$.

    (1) Monomial   (3) Trinomial

    (2) Binomial   (4) None of the above

2. Classify this polynomial $7x^2 - 3x + 2$.

    (1) Monomial   (3) Trinomial

    (2) Binomial   (4) None of the above

3. Multiply $3x^3 \cdot 4x^5$.

    (1) $7x^8$ (3) $12x^8$

    (2) $7x^{15}$ (4) $12x^{15}$

4. Which expression is equivalent to $2x^2 + 5x^2$?

    (1) $10x^2$

    (2) $7x^2$

    (3) $7x^4$

    (4) The expression cannot be simplified any further.

5. Simplify $6x(2x + 3)$.

    (1) $8x^2 + 18x$ (3) $12x^2 + 3$

    (2) $12x^2 + 18x$ (4) $20x$

6. Simplify $3x(5x^2 - 2x + 3)$.
   (1) $15x^3 - 2x + 3$ (3) $5x^2 + x + 3$
   (2) $15x^3 - 6x^2 + 9x$ (4) $15x^3 + 6x^2 - 9x$

7. Simplify $(3x - 4) - (5x - 3)$.
   (1) $-2x - 7$ (3) $2x - 1$
   (2) $-2x - 1$ (4) $2x - 7$

8. $(2x + 3)(3x - 1) =$
   (1) $6x^2 - 3$ (3) $6x^2 + 7x - 3$
   (2) $6x^2 + 11x - 3$ (4) $6x^2 - 7x - 3$

9. Factor $x^2 - 2x - 15$.
   (1) $(x - 3)(x - 5)$ (3) $(x - 15)(x + 1)$
   (2) $(x + 3)(x - 5)$ (4) $(x + 15)(x - 1)$

10. Factor $x^4 + 7x^2 + 12$.
    (1) $(x^2 + 6)(x^2 + 2)$
    (2) $(x^2 - 3)(x^2 - 4)$
    (3) $(x^2 + 3)(x^2 + 4)$
    (4) This cannot be factored.

Solutions

1. There are two terms, $5x^2$ and 3 separated by a $+$ sign. A polynomial with two terms is called a binomial.

   The correct choice is **(2)**.

2. There are three terms, $7x^2$, $3x$, and 2 separated by $+$ and $-$ signs. A polynomial with three terms is called a trinomial.

   The correct choice is **(3)**.

3. To multiply two monomials, first multiply the coefficients, $3 \cdot 4 = 12$. Then multiply the variable parts. Remember that when multiplying variables, you add the exponents. $x^3 \cdot x^5 = x^{(3+5)} = x^8$. The solution is $12x^8$.

   The correct choice is **(3)**.

4. Since these are like terms with variable part x^2, they can be combined. The solution will also have a variable part of x^2 with a coefficient equal to the sum of the two coefficients. Since $2 + 5 = 7$, $2x^2 + 5x^2 = 7x^2$.

   The correct choice is **(2)**.

5. Using the distributive property it becomes $6x \cdot 2x + 6x \cdot 3 = 12x^2 + 18x$.

   The correct choice is **(2)**.

6. Using the distributive property it becomes $3x \cdot 5x^2 + 3x(-2x) + 3x(3) = 15x^3 - 6x^2 + 9x$.

   The correct choice is **(2)**.

7. Distribute the $-$ sign through the parentheses on the right. The expression becomes $3x - 4 - 5x + 3$. Combine like terms to get $-2x - 1$.

   The correct choice is **(2)**.

8. Use the FOIL process. The firsts are $2x \cdot 3x = 6x^2$. The outers are $2x \cdot (-1) = -2x$. The inners are $3 \cdot 3x = 9x$. The lasts are $3 \cdot (-1) = -3$. Combine these four terms to get $6x^2 - 2x + 9x - 3$. Combine like terms to get $6x^2 + 7x - 3$.

   The correct choice is **(3)**.

9. To factor this quadratic trinomial, find two numbers that have a product of -15 and a sum of -2. The numbers are -5 and $+3$. The factors, then, are $(x - 5)(x + 3)$.

   The correct choice is **(2)**.

10. This trinomial can be written as $(x^2)^2 + 7(x^2) + 12$. It has the same structure, then, as a quadratic trinomial and can be factored by finding two numbers that have a product of 12 and a sum of 7. The two numbers are $+3$ and $+4$. The factors are $(x^2 + 3)$ and $(x^2 + 4)$.

    The correct choice is **(3)**.

3. QUADRATIC EQUATIONS

3.1 METHODS OF SOLVING QUADRATIC EQUATIONS

A **quadratic equation** is an equation that can be written in the form $ax^2 + bx + c = 0$. For example, $x^2 + 4x - 5 = 0$ is a quadratic equation. There are three ways to solve a quadratic equation.

- 1. Solve by factoring. If possible, factor the left-hand side of the equation.

$$x^2 + 4x - 5 = 0$$
$$(x + 5)(x - 1) = 0$$

Since the only way that two things can have a product of zero is if at least one of them is zero, this means that either $(x + 5)$ or $(x - 1)$ must equal zero.

$$
\begin{array}{ccc}
x + 5 = 0 & \text{or} & x - 1 = 0 \\
-5 = -5 & & +1 = +1 \\
x = -5 & \text{or} & x = 1
\end{array}
$$

$x = -5$ or $x = 1$ are the solutions to this quadratic equation.

- 2. Solve by completing the square. First eliminate the constant from the left-hand side by adding or subtracting.

$$x^2 + 4x - 5 = 0$$
$$+5 = +5$$
$$x^2 + 4x = 5$$

Next, *divide* the coefficient of the x by 2, square that answer, and add that number to both sides of the equation.

$$\left(\frac{4}{2}\right)^2 = 2^2 = 4$$
$$x^2 + 4x + 4 = 5 + 4$$
$$x^2 + 4x + 4 = 9$$

The left-hand side of the equation will factor.

$$(x + 2)(x + 2) = 9$$
$$(x + 2)^2 = 9$$

- Take the square root of both sides of the equation, putting a $\pm$ in front of the square root of the right-hand side.

$$\sqrt{(x+2)^2} = \pm\sqrt{9}$$
$$x + 2 = \pm 3$$
$$-2 = -2$$
$$x = -2 \pm 3$$
$$x = -2 + 3 \text{ or } x = -2 - 3$$
$$x = 1 \text{ or } x = -5$$

- 3. Solve with the quadratic formula. Any quadratic equation of the form $ax^2 + bx + c = 0$ can be solved with the equation

$$x = \frac{-b \pm \sqrt{b^2 - 4ac}}{2a}$$

For $x^2 + 4x - 5 = 0$, $a = 1, b = 4, c = -5$.

$$x = \frac{-4 \pm \sqrt{4^2 - 4 \cdot 1 \cdot (-5)}}{2 \cdot 1} = \frac{-4 \pm \sqrt{16 + 20}}{2} = \frac{-4 \pm \sqrt{36}}{2} = \frac{-4 \pm 6}{2}$$

$$x = \frac{-4 + 6}{2} = \frac{2}{2} = 1 \text{ or } x = \frac{-4 - 6}{2} = -\frac{10}{2} = -5$$

3.2 THE RELATIONSHIP BETWEEN FACTORS AND ZEROS

If a quadratic equation has factors $(x - p)$ and $(x - q)$ then the roots of the equation are p and q. If the equation has roots (or zeros) p and q, the factors are $(x - p)$ and $(x - q)$.

For example, the factors of the equation $x^2 - 7x + 10$ are $(x - 5)$ and $(x - 2)$. Therefore, the zeros of the equation are 5 and 2.

If the roots of a quadratic equation are -2 and 8, then the factors are $(x - (-2))$ and $(x - 8)$. The $(x - (-2))$ can be expressed as $(x + 2)$.

3.3 WORD PROBLEMS INVOLVING QUADRATIC EQUATIONS

Some real-world scenarios can be modeled with quadratic equations.

- **Area Problems**

The width of a rectangle is 3 units more than the length. If the area of the rectangle is 70 square units, what are the length and width of the rectangle?

$$A = 70 \quad | \quad l + 3$$

$$l$$

Since area is length times width, for this scenario

$$70 = l \cdot w$$
$$70 = l \cdot (l + 3)$$
$$70 = l^2 + 3l$$
$$0 = l^2 + 3l - 70$$

Any of the methods can be used to solve for the answers $l = -10$ and $l = 7$. Since the length must be positive, the answer is 7 units.

- **Projectile Problems**

The height of a projectile after t seconds can be modeled with a quadratic equation. If the equation for the height of a baseball is $h = -16t^2 + 48t + 64$, when will the baseball land on the ground?

When the ball lands on the ground, the height will be 0.
$0 = -16t^2 + 48t + 64$ is the equation.

- This can be solved by any of the methods for the answers $t = -1$ and $t = 4$. Since the amount of time must be positive, the solution is 4 seconds.

34

Practice Exercises

1. Find all solutions to $(x + 2)^2 = 64$.

    (1) $\sqrt{62}, -\sqrt{62}$ (3) 6

    (2) 6, −10   (4) −10

2. Use completing the square to find both solutions for x in the equation $x^2 + 8x + 16 = 9$.

    (1) −1, −7   (3) −2, −3

    (2) −1, −2   (4) −3, −4

3. Find all solutions to $x^2 + 6x = 0$.

    (1) 0   (3) 0, −6

    (2) −6   (4) 0, 6

4. Solve $x^2 + 10x + 24 = 0$ by factoring.

    (1) 4, −6   (3) −4, 6

    (2) −4, −6   (4) 4, 6

5. If the roots of an equation are 3 and −6, what could the equation be?

    (1) $(x - 3)(x - 6) = 0$ (3) $(x + 3)(x - 6) = 0$

    (2) $(x - 3)(x + 6) = 0$ (4) $(x + 3)(x + 6) = 0$

6. If the roots of a polynomial are 1 and -8, what could be the factors?
    (1) $(x - 1)$ and $(x + 8)$ (3) $(x + 1)$ and $(x + 8)$
    (2) $(x - 1)$ and $(x - 8)$ (4) $(x + 1)$ and $(x - 8)$

7. Solve $x^2 + 4x - 7 = 0$ with the quadratic formula.
    (1) 1.3 (3) $2 \pm \sqrt{12}$
    (2) $2 \pm \sqrt{11}$ (4) $-2 \pm \sqrt{11}$

8. The width of a rectangle is 10 inches longer than its length. If the area of the rectangle is 56 square inches, which equation could be used to determine its length (l)?
    (1) $l(l + 10) = 56$ (3) $2l + 2(l + 10) = 56$
    (2) $l(l - 10) = 56$ (4) $2l - 2(l + 10) = 56$

9. The height of a projectile in feet at time t is determined by the equation $h = -16t^2 + 128t + 320$. At what time will the projectile be 560 feet high?
    (1) 4 seconds   (3) 6 seconds
    (2) 5 seconds   (4) 7 seconds

10. The height of a projectile in feet at time t is determined by the equation $h = -16t^2 + 112t + 128$. At what time will the projectile be 0 feet high?
    (1) 5 seconds   (3) 7 seconds
    (2) 6 seconds   (4) 8 seconds

Solutions

1. Take the square root of both sides to get $x + 2 = \pm 8$. The equation $x + 2 = 8$ has solution $x = 6$. The equation $x + 2 = -8$ has solution $x = -10$.

 The correct choice is (**2**).

2. Since the constant, 16, is already equal to the square of half the coefficient $\left(\frac{8}{2}\right)^2$, the left-hand side of the equation is already a perfect square trinomial.

 It can be factored as $(x + 4)^2 = 9$. Then take the square root of both sides to get $x + 4 = \pm 3$. The equation $x + 4 = 3$ has solution $x = -1$. The equation $x + 4 = -3$ has solution $x = -7$.

 The correct choice is (**1**).

3. Factor out an x to get the equation $x(x + 6) = 0$. This equation is true when either $x = 0$ or when $x + 6 = 0$, which becomes $x = -6$.

 The correct choice is (**3**).

4. The two numbers that have a product of 24 and a sum of 10 are $+4$ and $+6$. The factors are $(x + 4)(x + 6)$. The solutions to the equation $(x + 4)(x + 6) = 0$ are when $x + 4 = 0$, which becomes $x = -4$ and also when $x + 6 = 0$, which becomes $x = -6$.

 The correct choice is (**2**).

5. When a is a root of an equation, $(x - a)$ is a factor. If the roots are 3 and -6, the factors can be $(x - 3)$ and $(x - (-6)) = (x + 6)$.

 The correct choice is (**2**).

6. If 1 is a root, $(x - 1)$ is a factor. If -8 is a root $(x - (-8)) = (x + 8)$ is a factor.

 The correct choice is (**1**).

7. $a = 1, b = 4, c = -7$.

$$x = \frac{-4 \pm \sqrt{4^2 - 4(1)(-7)}}{2} = \frac{-4 \pm \sqrt{16 + 28}}{2} = \frac{-4 \pm \sqrt{44}}{2}$$
$$= \frac{-4 \pm 2\sqrt{11}}{2} = -2 \pm \sqrt{11}$$

The correct choice is **(4)**.

8. The area of a rectangle is $l \cdot w$. If l is the length and $l + 10$ is the width, the area is $l(l + 10)$. If the area is known to be 56, the equation that can be used to solve for l is $l(l + 10) = 56$.

The correct choice is **(1)**.

9. When 560 is substituted for h, the equation becomes $560 = -16t^2 + 128t + 320$. Subtract 560 from both sides of the equation to get $0 = -16t^2 + 128t - 240$. Divide both sides by -16 to get $0 = t^2 - 8t + 15$. The right-hand side factors and the equation becomes $0 = (t - 3)(t - 5)$ with solutions $t = 3$ and $t = 5$. Of the choices listed, only 5 seconds is correct.

The correct choice is **(2)**.

10. Substituting $h = 0$ into the equation, it becomes $0 = -16t^2 + 112t + 128$. Divide both sides by -16 to get $0 = t^2 - 7t - 8$. The right side factors so the equation becomes $0 = (t - 8)(t + 1)$. The solutions are $t = 8$ and $t = -1$. Since the amount of time must be positive, the -1 is rejected.

The correct choice is **(4)**.

4. SYSTEMS OF LINEAR EQUATIONS

4.1 WHAT IS A SYSTEM OF LINEAR EQUATIONS?

A **system of equations** is a set of two equations that each have two variables. The system is *linear* if there are no exponents greater than one on any of the variables.

$$2x + 3y = 21$$
$$5x - 2y = 5$$

is a system of linear equations.

The solution to a system of equations is the set of **ordered pairs** that satisfy both equations at the same time. For the system above, the solution is the ordered pair $(3, 5)$ since

$$2(3) + 3(5) = 6 + 15 = 21$$
$$5(3) - 2(5) = 15 - 10 = 5$$

There are two main techniques for solving systems of equations.

4.2 SOLVING A SYSTEM OF EQUATIONS WITH THE SUBSTITUTION METHOD

If one of the two equations is in the form $y = mx + b$, use **the substitution method**.

$$y = 2x + 3$$
$$2x + 4y = 42$$

- Substitute the expression $2x + 3$ for the y in the second equation.

$$2x + 4(2x + 3) = 42$$

Solve for x.

$$2x + 8x + 12 = 42$$
$$10x + 12 = 42$$
$$-12 = -12$$
$$\frac{10x}{10} = \frac{30}{10}$$
$$x = 3$$

- Substitute 3 for x into either of the equations and solve for y.

$$y = 2(3) + 3$$
$$y = 6 + 3$$
$$y = 9$$

The solution is the ordered pair $(3, 9)$.

4.3 SOLVING A SYSTEM OF EQUATIONS WITH THE ELIMINATION METHOD

When both equations are in the form $ax + by = c$, use the **elimination method**. The elimination method is when the two equations are combined in a way that one of the variables is eliminated.

In the system of equations below, the y term will be eliminated if the two equations are added. When the coefficient of one of the variables in one equation is the same number with the opposite sign of the same variable in the other equation, add the equations to eliminate that variable. In this case, the $+2y$ has the opposite coefficient as the $-2y$.

$$3x + 2y = 19$$
$$+4x - 2y = 16$$
$$\overline{}$$
$$\frac{7x}{7} = \frac{35}{7}$$
$$x = 5$$

- Substitute 5 for x into either of the equations

$$3(5) + 2y = 19$$
$$15 + 2y = 19$$
$$-15 = -15$$
$$\frac{2y}{2} = \frac{4}{2}$$
$$y = 2$$

The solution is the ordered pair $(5, 2)$.

For some systems of equations, one or both of the equations need to be changed so the elimination will happen.

In the system

$$2x - 4y = 4$$
$$3x + 2y = 14$$

Since 4 is a multiple of 2, multiply both sides of the bottom equation by 2 and the coefficients of the y variables will be the same number with opposite signs.

$$2x - 4y = 4$$
$$2(3x + 2y) = 2(14)$$
$$2x - 4y = 4$$
$$+6x + 4y = 28$$
$$\frac{8x}{8} = \frac{32}{8}$$
$$x = 4$$

$$2(4) - 4y = 4$$
$$8 - 4y = 4$$
$$-8 = -8$$
$$\frac{-4y}{-4} = \frac{-4}{-4}$$
$$y = 1$$

The solution is the ordered pair (4, 1).

In the system

$$3x - 5y = 17$$
$$2x + 4y = 4$$

- To eliminate the y, determine the least common multiple of 4 and 5, which is 20. Multiply both sides of both equations so that one of the coefficients on the y is -20 and the other is $+20$. To do this, multiply both sides of the top equation by 4 and multiply both sides of the bottom equation by 5.

$$4(3x - 5y) = 4(17)$$
$$5(2x + 4y) = 5(4)$$

$$12x - 20y = 68$$
$$+10x + 20y = 20$$
$$\overline{}$$
$$\frac{22x}{22} = \frac{88}{22}$$
$$x = 4$$

- Substitute 4 for x in either of the original equations and solve for y.

$$3(4) - 5y = 17$$
$$12 - 5y = 17$$
$$-12 = -12$$
$$\frac{-5y}{-5} = \frac{5}{-5}$$
$$y = -1$$

The solution to the system of equations is the ordered pair $(4, -1)$.

4.4 WORD PROBLEMS INVOLVING SYSTEMS OF EQUATIONS

Some real-world scenarios can be modeled with a system of linear equations. Here is a typical example.

If five slices of pizza and three drinks cost \$21 and two slices of pizza and five drinks cost \$16, how much is it for just one slice of pizza?

- Let x be the cost of a slice of pizza and y be the cost of a drink.

The system of equations is

$$5x + 3y = 21$$
$$2x + 5y = 16$$

- To eliminate the y, make the $3y$ and the $5y$ into $15y$ and $-15y$, respectively, by multiplying both sides of the top equation by 5 and both sides of the bottom equation by -3. Then add the two equations, and solve for x.

$$5(5x + 3y) = 5(21)$$
$$-3(2x + 5y) = -3(16)$$

$$25x + 15y = 105$$
$$+ -6x - 15y = -48$$

$$\frac{19x}{19} = \frac{57}{19}$$
$$x = 3$$

Since the question just asked for the price of a slice of pizza, it is not necessary to also find the value of y. A slice of pizza costs \$3.

Practice Exercises

1. $(2, 5)$ is a solution to which equation?
   - (1) $x + 2y = 9$
   - (2) $2x + y = 9$
   - (3) $8x - y = 9$
   - (4) $3x + 4y = 25$

2. Which equation has the same solution set as the equation $2x + 3y = 5$?
   - (1) $8x + 12y = 20$
   - (2) $8x + 12y = 15$
   - (3) $6x + 9y = 12$
   - (4) $4x + 6y = 8$

3. Solve the system of equations

$$3x + 2y = 17$$
$$4x - 2y = 4$$

   - (1) $(4, 3)$
   - (2) $(5, 1)$
   - (3) $(3, 4)$
   - (4) $(1, 6)$

4. Solve the system of equations

$$y = 5x + 3$$
$$2x + 6y = 50$$

   - (1) $(1, 8)$
   - (2) $(8, 1)$
   - (3) $(-1, -8)$
   - (4) $(-8, -1)$

5. Solve the system of equations

$$2x + 3y = -1$$
$$-2x + 5y = -23$$

(1) $(-4, 3)$ (3) $(4, -3)$

(2) $(4, 3)$ (4) $(-4, -3)$

6. Solve the system of equations

$$y = 3x - 2$$
$$4x - 2y = -4$$

(1) $(10, 4)$ (3) $(-10, -4)$

(2) $(4, 10)$ (4) $(-4, -10)$

7. In order to eliminate the x from this system of equations,

$$12x - 3y = 21$$
$$-2x + 6y = 2$$

you could

(1) Multiply both sides of the first equation by 2.

(2) Multiply both sides of the second equation by 6.

(3) Multiply both sides of the first equation by -2.

(4) Multiply both sides of the second equation by 1/2.

8. Solve the system of equations

$$8x - 2y = 28$$
$$4x + 3y = 6$$

(1) $(3, 2)$ (3) $(-3, -2)$

(2) $(-3, 2)$ (4) $(3, -2)$

9. Which system of equations can be used to model the following scenario?

There are 50 animals. Some of the animals have two legs and the rest of them have four legs. In total there are 172 legs.

(1) $x + y = 172$ (3) $y + 50 = x$

     $2x + 4y = 50$ $4y + 172 = 2x$

(2) $x + 50 = y$ (4) $x + y = 50$

     $2x + 172 = 4y$ $2x + 4y = 172$

10. A pet store has 30 animals. Some are cats and the rest are dogs. The cats cost $50 each. The dogs cost $100 each. If the total cost for all 30 animals is $1,900, how many cats are there?

(1) 8   (3) 22

(2) 20   (4) 24

Solutions

1. Substitute 2 for x and 5 for y into each of the choices. Choice (1) becomes $2 + 2(5) = 9$, which is not true. Choice (2) becomes $2(2) + 5 = 9$, which is true.

    The correct choice is **(2)**.

2. If both sides of an equation are multiplied by the same number, the new equation has the same solution set as the original equation. If both sides of the equation $2x + 3y = 5$ are multiplied by 4, it becomes $8x + 12y = 20$, which is choice (1). The other choices could be obtained by multiplying the right-hand side of the original equation by one number and the left-hand side of the original equation by another number, which will not produce an equation with the same solution set as the original.

    The correct choice is **(1)**.

3. Since the top equation has a $+2y$ and the bottom equation has a $-2y$, the equations can be added together, and the y terms will drop out leading to the equation $7x = 21$ or $x = 3$. To solve for y, substitute 3 for x into either of the original equations, like $3(3) + 2y = 17$, $9 + 2y = 17$, $2y = 8$, which leads to $y = 4$.

    The correct choice is **(3)**.

4. Since the y is isolated in the top equation, substitute $5x + 3$ for y in the bottom equation. It becomes $2x + 6(5x + 3) = 50$, $2x + 30x + 18 = 50$, $32x + 18 = 50$, $32x = 32$, $x = 1$. Only one choice has an x-coordinate of 1.

    The correct choice is **(1)**.

5. Since the top equation has a $+2x$ and the bottom equation has a $-2x$, the equations can be added together to get $8y = -24$, $y = -3$. Substitute -3 for y into either equation to solve for x. $2x + 3(-3) = -1$, $2x - 9 = -1$, $2x = 8$, $x = 4$.

    The correct choice is **(3)**.

6. Since the y is isolated in the top equation, substitute $3x - 2$ for y in the bottom equation. $4x - 2(3x - 2) = -4$, $4x - 6x + 4 = -4$, $-2x + 4 = -4$, $-2x = -8$, $x = 4$. To solve for y, substitute 4 for x into either equation. $y = 3(4) - 2 = 12 - 2 = 10$.

The correct choice is **(2)**.

7. The x will be eliminated after combining two equations when the coefficient of the x in one of the equations is the opposite of the coefficient of the x in the other equation. For choice (1), if both sides of the top equation are multiplied by 2, it would become $24x - 6y = 42$. $+24$ is not the opposite of -2. For choice (2) if both sides of the second equation are multiplied by 6, it becomes $-12x + 36y = 12$. Since -12 is the opposite of $+12$, this is the best answer.

The correct choice is **(2)**.

8. Multiply both sides of the bottom equation by -2 to get $-8x - 6y = -12$. Add this to the top equation to eliminate the x and get $-8y = 16$. Divide both sides by -8 to get $y = -2$. Substitute -2 for y into one of the original equations. $8x - 2(-2) = 28$, $8x + 4 = 28$, $8x = 24$, $x = 3$.

The correct choice is **(4)**.

9. If x is the number of two-legged animals and y is the number of four-legged animals, the number of animals is $x + y$ and the number of legs is $2x + 4y$. The system, then, is $x + y = 50$ and $2x + 4y = 172$.

The correct choice is **(4)**.

10. If x is the number of dogs and y is the number of cats, the system of equations is

$$x + y = 30$$
$$100x + 50y = 1,900$$

To eliminate the x, multiply both sides of the top equation by -100 to get $-100x - 100y = -3,000$. Add this to the bottom equation to get $-50y = -1,100$ or $y = 22$.

The correct choice is **(3)**.

5. GRAPHS OF LINEAR EQUATIONS

5.1 GRAPHING THE SOLUTION SET OF A LINEAR EQUATION BY MAKING A TABLE OF VALUES

The equation $x + y = 10$ has an infinite number of ordered pairs that satisfy it. One way to organize the information before creating a graph is to make a table of values.

This is a chart with three ordered pairs satisfying the equation $x + y = 10$. For a linear equation, only two ordered pairs are needed, but it is wise to do an extra ordered pair in case one of your first two is incorrect.

x	y
2	8
3	7
9	1

Plot the ordered pair (2, 8) on the coordinate plane by locating the point that is two units to the right of the y-axis and 8 units above the x-axis. One way to do this is to start at the origin point where the two axes intersect and move to the right two units from there and then up 8 units.

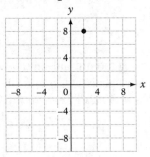

Do the same for the other two ordered pairs on the chart (3, 7) and (9, 1).

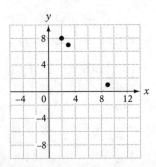

Draw a line through the three points. If the three points do not all lie on the same line, one of your ordered pairs is incorrect.

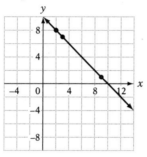

MATH FACTS

The line on a graph contains an infinite number of points. Each point corresponds to an ordered pair that is part of the solution set for the equation, and each ordered pair that is part of the solution set for the equation corresponds to a point on the line.

5.2 GRAPHING A LINEAR EQUATION USING THE INTERCEPT METHOD

When a linear equation is written in the form $ax + by = c$, the graph of the equation can be created quickly by finding the x-intercept and the y-intercept.

For the equation $2x - 3y = 12$:

- To find the y-intercept, substitute 0 for x and solve for y.

$$2(0) - 3y = 12$$
$$\frac{-3y}{-3} = \frac{12}{-3}$$
$$y = -4$$

The y-intercept is $(0, -4)$.

- To find the x-intercept, substitute 0 for y and solve for x.

$$2x - 3(0) = 12$$
$$2x = 12$$
$$x = 6$$

The x-intercept is $(6, 0)$.

- Plot $(0, -4)$ and $(6, 0)$ on a set of coordinate axes and draw the line that passes through both points.

5.3 CALCULATING AND INTERPRETING SLOPE

The *slope* of a line is a number that measures how steep it is. A horizontal line has a slope of 0. A line with a positive slope goes up as it goes to the right. A line with a negative slope goes down as it goes to the right. The variable used for slope is the letter m.

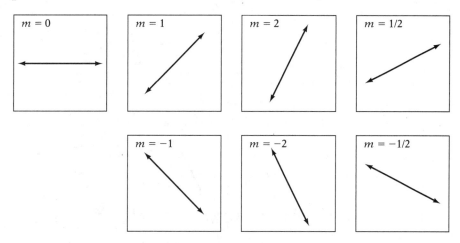

5.4 GRAPHING A LINEAR EQUATION IN SLOPE-INTERCEPT FORM

When the linear equation is in the form $y = mx + b$, it can be graphed very quickly. The y-intercept is $(0, b)$, and from that point to another point move one unit to the right and m units up.

For the equation $y = 2x - 3$, the y-intercept is $(0, -3)$. From that point, move 1 unit to the right and 2 units up. Draw the line through the two points $(0, -3)$ and $(1, -1)$.

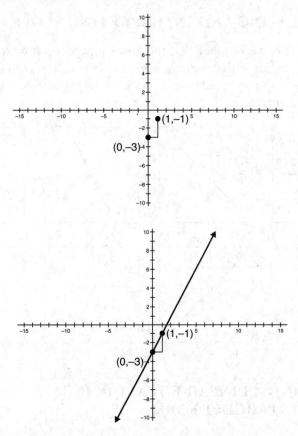

- If m is a fraction, n/d, it is more accurate to move d units to the right and n units up from the y-intercept.

- For the equation $y = \dfrac{2}{3}x + 2$, the y-intercept is $(0, 2)$. From that point move 3 to the right and 2 up to get to the point $(3, 4)$. Draw a line through these two points.

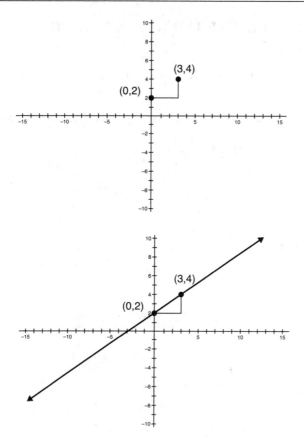

If the *m* value is negative, move down instead of up to get from the *y*-intercept to the next point.

5.5 GRAPHING LINEAR EQUATIONS ON THE GRAPHING CALCULATOR

An equation in slope-intercept form can be quickly graphed on the graphing calculator.

- For the TI-84, press [Y=], enter the equation, and press [ZOOM] and [6].

- For the TI-Nspire, go to the home screen and select [B] for the Graph Scratchpad. Enter the equation on the entry line after $f1(x)=$ and press [enter].

- If a linear equation with two variables is not originally in the slope-intercept form, $y = mx + b$, algebra can be used to rewrite it in slope-intercept form before graphing.

5.6 EQUATIONS OF VERTICAL AND HORIZONTAL LINES

The graph of the equation $y = k$ is a horizontal line through the point $(0, k)$.
The graph of the equation $x = h$ is a vertical line through the point $(h, 0)$.

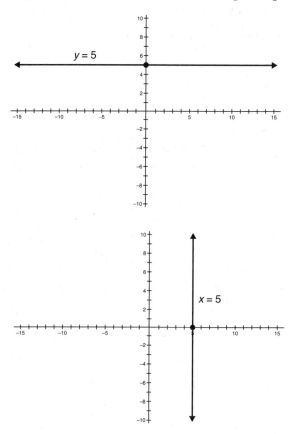

5.7 SOLVING SYSTEMS OF LINEAR EQUATIONS GRAPHICALLY

The solution to a system of equations is the set of coordinates of the point where the graphs of the two lines intersect.

- The solution to the system of equations

$$y = 2x - 1$$

$$y = -\frac{2}{3}x + 7$$

can be found by carefully producing the two graphs on the same set of axes. The coordinates of the intersection point $(3, 5)$ means the solution to the system of equations is $x = 3$, $y = 5$.

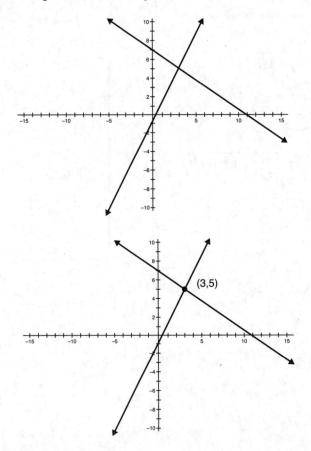

The graphing calculator can also determine the intersection point of two lines. On the TI-84, graph both lines and press [2ND], [TRACE], and [5] to find the intersection. On the TI-Nspire, graph both lines and press [menu], [6], and [4] to find the intersection.

5.8 DETERMINING THE SLOPE OF A LINE

If two points on a line are (x_1, y_1) and (x_2, y_2), then the slope of the line can be determined by the formula

$$m = \frac{y_2 - y_1}{x_2 - x_1}$$

The slope of the line passing through $(3, 5)$ and $(7, 8)$ is

$$m = \frac{8 - 5}{7 - 3} = \frac{3}{4}$$

5.9 INTERPRETING THE SLOPE OF A LINE

When a graph represents a real-world scenario, the slope is the rate that the x value is changing with relation to the y value. In a distance-time graph, the slope of a line segment corresponds to the speed that the moving object is traveling.

In the distance-time graph above, a car's time traveled is represented by the x-coordinate, and the distance it has covered at that time is represented by the y-coordinate. The slope of the segment connecting $(0, 0)$ and $(5, 300)$ is 60 so the car was traveling at a speed of 60 mph for the first 5 hours. The slope of the segment connecting $(5, 300)$ and $(8, 540)$ is 80 so the car was traveling at a speed of 80 mph for the last 3 hours.

5.10 FINDING THE EQUATION OF A LINE THROUGH TWO GIVEN POINTS

If two points are known, use the slope-intercept formula to find the slope of the line. Substitute the value you calculated for m and the x- and y-coordinates of either of the given two points into the equation $y = mx + b$ and solve for b. Substitute the values calculated for m and b into the equation $y = mx + b$.

- For the line through the two points $(3, 5)$ and $(9, 1)$, m is the slope of the

  line $= \dfrac{1 - 5}{9 - 3} = -\dfrac{4}{6} = -\dfrac{2}{3}$.

- Using the point $(3, 5)$, substitute $x = 3$, $y = 5$, $m = -\dfrac{2}{3}$ into the equation $y = mx + b$ and solve for b.

$$5 = -\frac{2}{3}(3) + b$$
$$5 = -2 + b$$
$$+2 = +2$$
$$7 = b$$

$m = -\dfrac{2}{3}$ and $b = 7$ so the equation is $y = -\dfrac{2}{3}x + 7$.

Practice Exercises

1. What is the x-intercept of the graph of the solution set of the equation $2x + 5y = 20$?

   (1) $(10, 0)$ (3) $(0, 10)$

   (2) $(4, 0)$ (4) $(0, 4)$

2. This is a graph of the solution set of which equation?

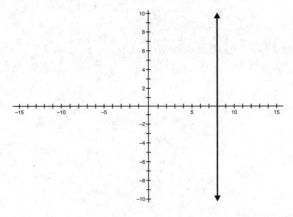

   (1) $x = 8$ (3) $y = 8$

   (2) $x = -8$ (4) $y = -8$

3. Below is the graph of the solution set of an equation. Based on this graph, which ordered pair does not seem to be part of the solution set of the equation $y = \frac{1}{3}x + 2$?

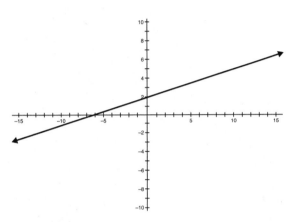

(1) $(3, 3)$ (3) $(6, 8)$

(2) $(6, 4)$ (4) $(9, 5)$

4. What is the slope of the line that passes through $(-2, 1)$ and $(8, 5)$?

(1) $\dfrac{2}{5}$ (3) $\dfrac{5}{2}$

(2) $-\dfrac{2}{5}$ (4) $-\dfrac{5}{2}$

5. Below is a distance-time graph for a bicycle trip. During which time interval is the cyclist going the fastest?

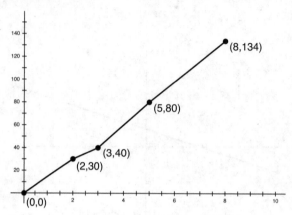

(1) 0 to 2 hours   (3) 3 to 5 hours

(2) 2 to 3 hours   (4) 5 to 8 hours

6. What is the slope of the line defined by the equation $y = -3x + 4$?

(1) 3   (3) 4

(2) −3   (4) −4

7. This is the graph of the solution set of which equation?

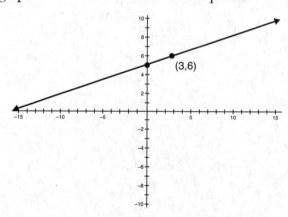

(1) $y = 5x + \dfrac{1}{3}$ (3) $y = 3x + 5$

(2) $y = 5x + 3$ (4) $y = \dfrac{1}{3}x + 5$

8. Find the equation of the line passing through the two points $(0, -7)$ and $(5, 8)$.

(1) $y = 3x + 7$ (3) $y = \dfrac{1}{3}x + 7$

(2) $y = 3x - 7$ (4) $y = \dfrac{1}{3}x - 7$

9. Find the equation of the line passing through the two points $(4, -2)$ and $(12, 4)$.

(1) $y = \dfrac{4}{3}x - 5$ (3) $y = \dfrac{3}{4}x - 5$

(2) $y = \dfrac{4}{3}x + 5$ (4) $y = \dfrac{3}{4}x + 5$

10. Find the equation of the line through the points $(3, 5)$ and $(3, 8)$.

(1) $x = -3$ (3) $y = -3$

(2) $x = 3$ (4) $y = 3$

Solutions

1. The x-intercept has a y-coordinate of zero. Substitute zero for y to get $2x + 5(0) = 20$, $2x = 20$, $x = 10$. The x-intercept is $(10, 0)$.

   The correct choice is **(1)**.

2. The points on this vertical line all have x-coordinates of 8. The equation is $x = 8$.

   The correct choice is **(1)**.

3. When the four points are plotted on the same graph as the line, only $(6, 8)$ is not on the line.

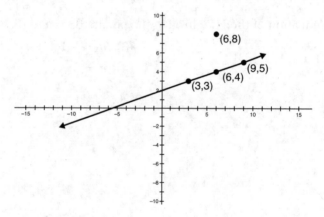

   The correct choice is **(3)**.

4. Using the slope formula $m = \dfrac{y_2 - y_1}{x_2 - x_1}$, with the points $(-2, 1)$ and $(8, 5)$ becomes $m = \dfrac{5 - 1}{8 - (-2)} = \dfrac{4}{10} = \dfrac{2}{5}$.

   The correct choice is **(1)**.

5. The interval between 0 and 2 has a slope of $30/2 = 15$. The interval between 2 and 3 has a slope of $10/1 = 10$. The interval between 3 and 5 has a slope of $40/2 = 20$. The interval between 5 and 8 has a slope of $54/3 = 18$. Since the slope represents the speed the bicycle is going, the interval between 3 and 5 hours is the fastest. It can also be seen from the graph that the interval between 3 and 5 hours looks the steepest.

   The correct choice is (**3**).

6. When the equation for a line is in $y = mx + b$ form, the m is the slope. For this equation, the coefficient of the x is -3 so the slope of the line is -3.

   The correct choice is (**2**).

7. According to the graph, the y-intercept is $(0, 5)$. The slope of the line through $(0, 5)$ and $(3, 6)$ is $m = \dfrac{6 - 5}{3 - 0} = \dfrac{1}{3}$. In $y = mx + b$ form, the m is the slope and the b is the y-coordinate of the y-intercept so the equation is $y = \dfrac{1}{3}x + 5$.

   The correct choice is (**4**).

8. The slope of the line is $m = \dfrac{8 - (-7)}{5 - 0} = \dfrac{15}{5} = 3$. Choose one of the points for the x and the y value and 3 for the m value. Using the point $(5, 8)$ the equation $y = mx + b$ becomes $8 = 3(5) + b$, $8 = 15 + b$, $b = -7$. The equation is $y = 3x - 7$.

   The correct choice is (**2**).

9. The slope of the line is $m = \dfrac{4 - (-2)}{12 - 4} = \dfrac{6}{8} = \dfrac{3}{4}$. Pick one of the points for x and y and substitute $\dfrac{3}{4}$ for m into $y = mx + b \cdot 4 = \dfrac{3}{4}$. $12 + b$, $4 = 9 + b$, $b = -5$. The equation is $y = \dfrac{3}{4}x - 5$.

   The correct choice is (**3**).

10. Since the x-coordinates are the same, the line is a vertical line. Vertical lines have equations $x = $ constant. Since the x-coordinates are both 3, the equation is $x = 3$.

    The correct choice is (**2**).

6. GRAPHS OF QUADRATIC EQUATIONS

6.1 GRAPHING A QUADRATIC EQUATION WITH A CHART

A **quadratic equation** can be written in the form $y = ax^2 + bx + c$. The graph of a quadratic equation is always a **parabola**, which resembles the letter U. When the a value is positive, the parabola looks like a right-side-up U. When the a value is negative, the parabola looks like an upside-down U.

- The simplest way to create the graph of a quadratic equation is to choose at least five consecutive x values and create a chart.

For the equation $y = x^2 - 2x - 3$, the chart could look like this:

x	y
-2	$(-2)^2 - 2(-2) - 3 = 4 + 4 - 3 = 5$
-1	$(-1)^2 - 2(-1) - 3 = 1 + 2 - 3 = 0$
0	$(0)^2 - 2(0) - 3 = -3$
1	$(1)^2 - 2(1) - 3 = 1 - 2 - 3 = -4$
2	$(2)^2 - 2(2) - 3 = 4 - 4 - 3 = -3$

- The graph of these five ordered pairs looks like this:

- *Connect* the points with a U-shaped parabola.

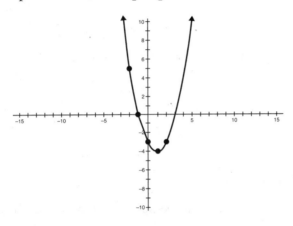

The low point of the parabola is called the **vertex**. In this example the vertex is $(1, -4)$. The two x-intercepts of the parabola are $(-1, 0)$ and $(3, 0)$. The y-intercept of this parabola is $(0, -3)$.

6.2 GRAPHING A PARABOLA BY FINDING THE VERTEX AND INTERCEPTS

The coordinates of the vertex, the y-intercept, and the x-intercepts can be calculated. These four points help produce an accurate graph of the parabola.

- The y-intercept of the parabola for the equation $y = ax^2 + bx + c$ is $(0, c)$.
- For the equation $y = x^2 - 2x - 3$, the y-intercept is $(0, -3)$.

The x-coordinate of the vertex of the parabola for the equation $y = ax^2 + bx + c$ is $x = -\dfrac{b}{2a}$. To get the y-coordinate, substitute the $-\dfrac{b}{2a}$ value in for x in the equation $y = ax^2 + bx + c$ to solve for y.

- For the equation $y = x^2 - 2x - 3$, the x-coordinate of the vertex is $x = -\dfrac{-2}{2(1)} = \dfrac{2}{2} = 1$. The y-coordinate is $y = 1^2 - 2(1) - 3 = 1 - 2 - 3 = -4$.
  So the vertex is $(1, -4)$.

The x-intercepts of the parabola for the equation $y = ax^2 + bx + c$ can be found by solving the quadratic equation $0 = ax^2 + bx + c$.

For the equation $y = x^2 - 2x - 3$, this becomes

$$0 = x^2 - 2x - 3$$

This quadratic equation can be solved by factoring.

$$0 = (x + 1)(x - 3)$$
$$x + 1 = 0 \text{ or } x - 3 = 0$$
$$-1 = -1 \quad\quad + 3 = +3$$
$$x = -1 \quad \text{or} \quad x = 3$$

The x-intercepts are $(-1, 0)$ and $(3, 0)$.

These four points help make an accurate sketch of the parabola.

6.3 GRAPHING QUADRATIC EQUATIONS ON THE GRAPHING CALCULATOR

The graphing calculator can easily graph quadratic equations. On the TI-84 press [Y=] and enter the equation after "Y1=", and press [ZOOM] and [6]. On the TI-Nspire, from the home screen press [B] for the Graph Scratchpad. Then enter the equation on the entry line and press [enter].

The graphing calculator can also determine the vertex and the intercepts using the min/max feature and the zeros feature.

- For the TI-84, press [2ND], [TRACE], and [3].

- For the TI-Nspire, press [menu], [6], and [2] to find the minimum point.

6.4 USING THE GRAPHING CALCULATOR TO SOLVE QUADRATIC EQUATIONS

The solutions to the equation $ax^2 + bx + c = 0$ are also the x-coordinates of the x-intercepts of the parabola defined by $y = ax^2 + bx + c$.

- To solve the equation $x^2 - 2x - 3 = 0$ with the graphing calculator, graph $y = x^2 - 2x - 3$ and then use the zeros feature to find the x-intercepts.
- For the TI-84, enter the equation $Y1 = x^2 - 2x - 3$ and [ZOOM] and [6] to graph. Then press [2ND], [TRACE], and [2] to find each zero.

- For the TI-Nspire, enter the equation $f1(x) = x^2 - 6x + 8$ to graph it. Then press [menu], [6], and [1] to find each zero.

6.5 SOLVING A LINEAR-QUADRATIC SYSTEM OF EQUATIONS BY GRAPHING

When a system of equations has one linear equation and one quadratic equation, one way to find the solution is to graph the line and the parabola and find the coordinates of the intersection point, or intersection points. There can be up to two solutions.

- **For the System of Equations**

$$y = -2x + 1$$
$$y = x^2 - 2x - 3$$

- Create the graph for both equations either by hand or with the graphing calculator. The coordinates of the two intersection points are the two solutions to the system of equations. For this system the two solutions are $x = -2, y = 5$ and $x - 2, y = -3$.

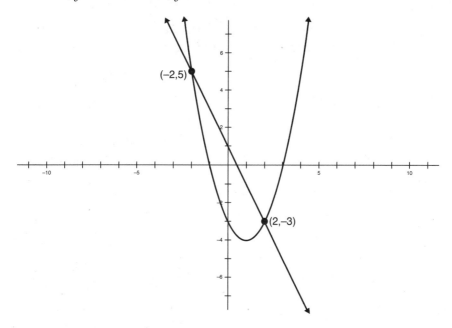

This can also be done on the graphing calculator.

- For the TI-84, graph the two equations and press [2ND], [TRACE], and [5] for each intersection point.

- For the TI-Nspire, graph the two equations and press [menu], [6], and [4] to find the intersection points.

6.6 WORD PROBLEMS INVOLVING THE GRAPH OF A QUADRATIC EQUATION

The height of a projectile thrown in the air can be modeled with a quadratic equation. The graph of this equation provides information about when the projectile reaches its highest point and when the projectile lands on the ground.

- If the equation is $h = -16t^2 + 48t + 160$, the y-coordinate of the vertex of the graph is the highest point the projectile reaches. The x-coordinate of the vertex is the amount of time it takes for the projectile to reach its highest point. The x-coordinate of the x-intercept is the amount of time it takes for the projectile to land. For this example, the highest point the projectile reaches is 196 feet high after 1.5 seconds. The projectile lands after 5 seconds.

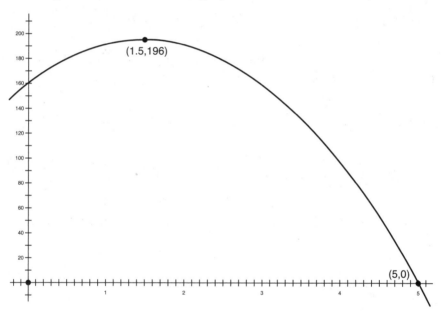

Practice Exercises

1. Which is a point on the graph of the solution set of $y = x^2 + 5x - 2$?

(1) $(3, 19)$

(2) $(3, 20)$

(3) $(3, 21)$

(4) $(3, 22)$

2. What are the coordinates of the vertex of the parabola defined by the equation $y = x^2 - 4x - 1$?

(1) $(-2, 5)$

(2) $(2, 5)$

(3) $(-2, -5)$

(4) $(2, -5)$

3. $x = -4$ is the x-coordinate of the vertex for the parabola defined by which equation?

(1) $y = x^2 + 8x + 3$

(2) $y = x^2 - 8x + 3$

(3) $y = x^2 + 4x + 3$

(4) $y = x^2 - 4x + 3$

4. What could be the equation that determines this parabola?

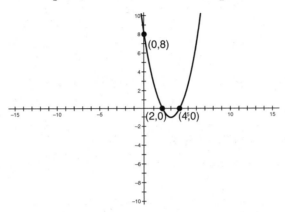

(1) $y = x^2 - 6x - 8$ (3) $y = x^2 + 6x - 8$

(2) $y = x^2 - 6x + 8$ (4) $y = x^2 + 6x + 8$

5. Which is the graph of $y = -x^2 + 2x + 3$?

(1)

(3)

(2)

(4)

6. Which ordered pair is a solution to the system?

$$y = x^2$$
$$y = x + 2$$

(1) $(1, 4)$ (3) $(3, 4)$

(2) $(2, 4)$ (4) $(4, 4)$

7. Solve this system of equations using algebra.

$$y = x^2$$
$$y = 2x + 3$$

(1) $(-1, 1)$ and $(9, 3)$ (3) $(-1, 1)$ and $(-3, 9)$

(2) $(-1, 1)$ and $(3, 9)$ (4) $(1, -1)$ and $(9, 3)$

8. Which system of equations could be used to solve this graph?

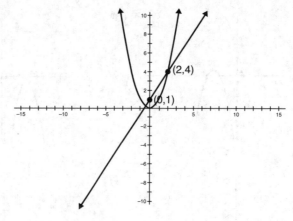

(1) $y = x^2$

$$y = \frac{2}{3}x + 1$$

(3) $y = x^2$

$$y = \frac{3}{2}x + 1$$

(2) $y = x^2$

$$y = \frac{3}{2}x - 1$$

(4) $y = x^2$

$$y = \frac{2}{3}x - 1$$

9. The x-intercepts of the parabola defined by which equation are the solutions to the equation $x^2 + 5x = 15$?

(1) $y = x^2 + 5x + 15$

(3) $y = x^2 + 5x - 15$

(2) $y = x^2 + 5x$

(4) $y = x^2 - 5x - 15$

10. Solve for all values of x, rounded to the nearest hundredth, $x^2 + 10x + 23 = 0$.

(1) $-3.41, -6.58$

(3) $-3.31, -6.64$

(2) $-3.59, -6.41$

(4) $-3.62, -6.18$

Solutions

1. For each choice, substitute the x-coordinate in for x and the y-coordinate in for y to see which makes the equation true. 22 does equal $3^2 + 5(3) - 2$, so the solution is $(3, 22)$.

   The correct choice is (**4**).

2. The x-coordinate of the vertex of a parabola $y = ax^2 + bx + c$ is $x = \dfrac{-b}{2a}$.

   Since a is 1 and b is -4, the x-coordinate of the vertex is $x = -\dfrac{(-4)}{2(1)} = 2$. To get the y-coordinate of the vertex, substitute 2 for x into the equation and solve for y. $y = 2^2 - 4(2) - 1 = 4 - 8 - 1 = -5$. The vertex is $(2, -5)$.

   The correct choice is (**4**).

3. Graph each choice on the graphing calculator and use the minimum feature to find the vertex of each. The vertex of the parabola $y = x^2 + 8x + 3$ is $(-4, -15)$, which has an x-coordinate of -4.

   The correct choice is (**1**).

4. Since the x-intercepts are $(2, 0)$ and $(4, 0)$, the equation for the parabola is $y = a(x - 2)(x - 4) = a(x^2 - 6x + 8)$. Since the y-intercept is $+8$, $a = 1$ and the equation is $y = x^2 - 6x + 8$.

   The correct choice is (**2**).

5. The x-coordinate of the vertex must be $x = -\dfrac{2}{2(1)} = 1$.

   The correct choice is (**2**).

6. Substitute x^2 for y in the bottom equation to get $x^2 = x + 2$, $x^2 - x - 2 = 0$, $(x - 2)(x + 1) = 0$, $x = 2$ or $x = -1$. The x-coordinates of the intersection points are 2 and -1. This can also be done with the graphing calculator. The answer is $(2, 4)$.

   The correct choice is (**2**).

7. Substitute x^2 for y in the bottom equation to get $x^2 = 2x + 3$, $x^2 - 2x - 3 = 0$, $(x - 3)(x + 1) = 0$. $x = 3$ and $x = -1$ are the x-coordinates of the two intersection points.

   The correct choice is **(2)**.

8. The parabola with vertex $(0, 0)$ passing through $(2, 4)$ is $y = x^2$. The line through $(0, 1)$ and $(2, 4)$ is $y = (3/2)x + 1$.

   The correct choice is **(3)**.

9. This equation can be rewritten as $x^2 + 5x - 15 = 0$. Equations in this form can be solved by finding the x-intercepts of the parabola $y = x^2 + 5x - 15$.

   The correct choice is **(3)**.

10. Using the graphing calculator, graph $y = x^2 + 10x + 23$ and use the "zero" feature to get $x = 3.59$ or $x = -6.41$.

    The correct choice is **(2)**.

7. LINEAR INEQUALITIES

7.1 ONE-VARIABLE LINEAR INEQUALITIES

A linear inequality is like a linear equation, but instead of an $=$ sign there is either a $>$, $<$, $\geq$, or $\geq$ sign. Solving a one-variable linear inequality is almost the same as solving a linear equality. The only difference is that when multiplying or dividing both sides by a negative number to eliminate the coefficient, the direction of the inequality sign must be reversed.

$$-2x + 3 < 11$$

- Eliminate the constant 3 by subtracting it from both sides of the inequality.

$$-2x + 3 < 11$$
$$-3 = -3$$
$$-2x < 8$$

- *Divide* both sides by -2 to isolate the x. Because you are dividing by a negative, the direction of the inequality sign must be reversed to keep the equation true. If the -2 were a $+2$, the direction of the inequality sign would not need to be reversed.

$$\frac{-2x}{-2} < \frac{8}{-2}$$
$$x > -4$$

The solution is $x > -4$.

7.2 GRAPHING TWO-VARIABLE INEQUALITIES

$y \leq 2x - 5$ is a two-variable inequality. To graph the two-variable inequality, first graph the line $y = 2x - 5$.

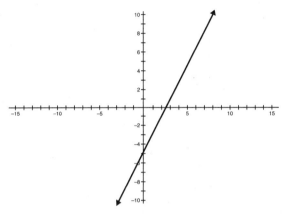

One side of the line needs to be shaded in. To determine which side of the line to shade, substitute the ordered pair $(0, 0)$ into the inequality to test if it yields a true inequality.

$$0 < 2(0) - 5$$
$$0 < 0 - 5$$
$$0 < -5 \text{ is not true.}$$

Since $(0, 0)$ does not satisfy the inequality, shade the side of the line that does not contain $(0, 0)$.

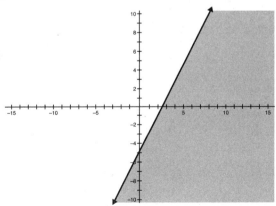

- If the inequality sign is a < or a >, the line must be a dotted line. If it is a ≤ or a ≥, it must be a solid line. If $(0, 0)$ is on the line, an ordered pair that is not on the line must be used to test which side to shade.
- For the graph of $y > 3x$, graph the line $y = 3x$ and make it a dotted line.

Since $(0, 0)$ is on the line, test a point that is not on the line, like $(2, 0)$.

$$0 > 3(2)$$

Since $0 > 6$ is not true, shade the side that does not contain $(2, 0)$.

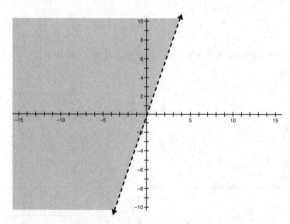

7.3 GRAPHING SYSTEMS OF LINEAR INEQUALITIES

To graph the solution set to a system of linear inequalities, graph both inequalities on the same graph and locate the portion of the graph that is shaded twice.

$$y > 2x - 5$$

$$y < -\frac{2}{3}x + 2$$

The portion of the graph that has an S is the solution set. Any point in that region will satisfy the system of inequalities. $(-3, 1)$ is an example of a point that is in this double-shaded region and will satisfy both inequalities.

7.4 GRAPHING INEQUALITIES ON THE GRAPHING CALCULATOR

The TI-84 and the TI-Nspire c an graph linear inequalities.

- On the TI-84, the shading can be set by moving the cursor to the icon to the left of the Y= and pressing [ENTER] until the icon shows the proper shading.

- On the TI-Nspire, delete the "=" from the equation in the entry line and choose the symbol to replace it with.

The graphing calculator can also graph systems of linear inequalities by graphing both inequalities on the same set of coordinate axes.

Practice Exercises

1. What is the solution set for $3x < -18$?

   (1) $x < -6$

   (2) $x > -6$

   (3) $x \geq -6$

   (4) $x \leq -6$

2. What is the solution set for $-4x \geq 20$?

   (1) $x \geq -5$

   (2) $x > -5$

   (3) $x \leq -5$

   (4) $x < -5$

3. What is the smallest integer that satisfies the equation $-6x < -18$?

   (1) 3

   (2) 4

   (3) 5

   (4) 6

4. Which is the graph of $y < x + 4$?

(1)

(3)

(2)

(4)

5. Below is the graph of $x + y \leq 8$. Which is a point in the solution set?

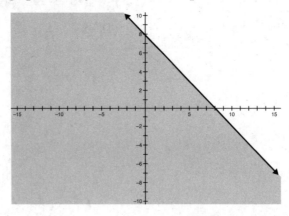

(1) $(1, 8)$ (3) $(4, 5)$

(2) $(2, 7)$ (4) $(3, 5)$

6. All of these ordered pairs are part of the shaded region for the graph of $2x + y \leq 12$ except

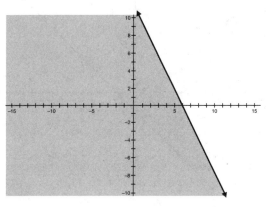

(1) $(2, 8)$ (3) $(5, 3)$

(2) $(3, 6)$ (4) $(4, 4)$

7. Which graph shows the solution to the system of inequalities?

$$y < 2x + 1$$

$$y > \frac{1}{3}x + 4$$

(1)

(3)

(2)

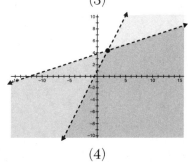

(4)

8. Which system of inequalities does the following graph show the solution for?

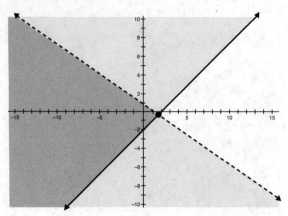

(1) $y \geq x - 2$

    $y < -\frac{2}{3}x + 1$

(2) $y \leq x - 2$

    $y > -\frac{2}{3}x + 1$

(3) $y < x - 2$

    $y \geq -\frac{2}{3}x + 1$

(4) $y > x - 2$

    $y \leq -\frac{2}{3}x + 1$

9. Which graph has the solution set shaded in for the following system of inequalities?

$$y \leq -x + 6$$

$$y \geq \frac{1}{2}x - 1$$

(1)

(2)

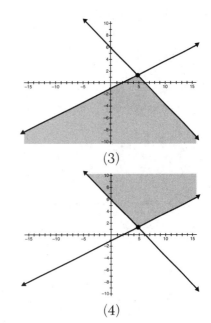

(3)

(4)

10. Which is the graph of the following system of inequalities?

$$y \geq 0$$
$$x \leq 0$$

(1)

(3)

(2)

(4)

Solutions

1. Divide both sides of the inequality by 3 to get $x < -6$.

   The correct choice is (**1**).

2. Divide both sides of the inequality by -4. Switch the direction of the inequality sign because you divided by a negative to get $x \leq -5$.

   The correct choice is (**3**).

3. Divide both sides of the inequality by -6 and switch the direction of the inequality sign to get $x > 3$. The smallest integer that satisfies the equation is 4. The answer is not 3 since it is not true that $3 > 3$.

   The correct choice is (**2**).

4. Because it is a $<$ and not a $\leq$ sign, the line must be dotted. Test to see if $(0, 0)$ is in the solution set by checking if $0 < 0 + 4$ is true. Since it is true, the side of the line containing $(0, 0)$ must be shaded.

   The correct choice is (**4**).

5. When all four choices are plotted on the graph, three of them are not in the shaded region. The point $(3, 5)$ is on the line, but since there is a $\leq$ sign, the line is part of the solution set.

   The correct choice is (**4**).

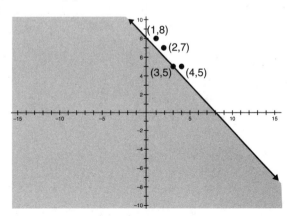

6. Three of the points are on the line, which is part of the solution set because it is a $\leq$ sign. The point $(5, 3)$ is not in the shaded area.

The correct choice is (**3**).

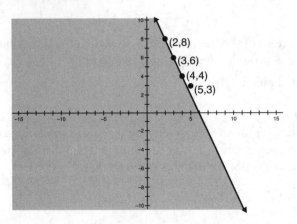

7. Both of the lines have to be dotted because of the inequality signs $<$ and $>$. This eliminates choices 1 and 3. Both choices 2 and 4 have the region below the line $y = 2x + 1$ shaded. The difference between choices 2 and 4 is that choice 2 has the region above $y = (1/3)x + 4$ shaded, and choice 4 has the region below $y = (1/3)x + 4$ shaded. To check which side is correct, substitute $(0,0)$ into the equation $y > (1/3)x + 4$ to get $0 > (1/3)(0) + 4$. Since 0 is not greater than 4, $(0, 0)$ is not part of the solution set to $y > (1/3)x + 4$. The side of the line $y = (1/3)x + 4$ that contains $(0, 0)$ should not be shaded so the region above the line $y = (1/3)x + 4$ should be shaded.

The correct choice is (**2**).

8. Since the line $y = x - 2$ is solid, choices 3 and 4 can be eliminated. Substituting $(0, 0)$ into both inequalities in choice 1 makes them both true while substituting $(0, 0)$ into both inequalities in choice 2 makes them both false.

   The correct choice is (**1**).

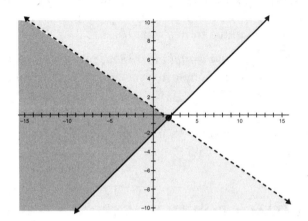

9. To check choice 1, pick a point in the shaded region for that choice and check to see if it satisfies both inequalities. Since $(0, 0)$ is a point in the shaded region, check to see if $0 \leq -0 + 6$ and $0 \geq (1/2) \cdot 0 - 1$. $(0, 0)$ does make both inequalities true.

   The correct choice is (**1**).

10. The points on or above the x-axis make $y \geq 0$ true. The points on or to the left of the y-axis make $x \leq 0$ true. The points that satisfy both inequalities must be above the x-axis and to the left of the y-axis.

    The correct choice is (**2**).

8. EXPONENTIAL EQUATIONS

8.1 EVALUATING EXPONENTIAL EXPRESSIONS

An **exponential equation** is one where one of the variables is an exponent. The equation $y = 3 \cdot 2^x$ is a two-variable exponential equation. Exponential equations can be written in the form $y = a \cdot b^x$.

- When evaluating an *exponential expression*, raise the base to the exponent before multiplying. For example, to evaluate $y = 3 \cdot 2^x$ when $x = 4$ it becomes

$$y = 3 \cdot 2^4$$
$$y = 3 \cdot 16 \ (\text{NOT } y = 6^4)$$
$$y = 48$$

8.2 EXPONENTIAL GROWTH VS. EXPONENTIAL DECAY

In the exponential equation $y = a \cdot b^x$, the b is called the **base**. When b is greater than 1, the equation is an example of *exponential growth* since $a \cdot b^x$ grows as x grows. When b is between 0 and 1 the equation is an example of *exponential decay* since $a \cdot b^x$ gets smaller as x grows.

$y = 3 \cdot 2^x$ is an example of exponential growth since $b = 2$, which is greater than 1.

$y = 3 \cdot (0.7)^x$ is an example of exponential decay since $b = 0.7$, which is between 0 and 1.

8.3 GRAPHS OF EXPONENTIAL EQUATIONS

A table of values for a two-variable exponential equation, like $y = 3 \cdot 2^x$, looks like this:

x	y
-3	$3 \cdot 2^{-3} = 3 \cdot \dfrac{1}{2^3} = 3 \cdot \dfrac{1}{8} = \dfrac{3}{8}$
-2	$3 \cdot 2^{-2} = 3 \cdot \dfrac{1}{2^2} = 3 \cdot \dfrac{1}{4} = \dfrac{3}{4}$
-1	$3 \cdot 2^{-1} = 3 \cdot \dfrac{1}{2^1} = 3 \cdot \dfrac{1}{2} = \dfrac{3}{2}$
0	$3 \cdot 2^0 = 3 \cdot 1 = 3$
1	$3 \cdot 2^1 = 3 \cdot 2 = 6$
2	$3 \cdot 2^2 = 3 \cdot 4 = 12$
3	$3 \cdot 2^3 = 3 \cdot 8 = 24$

Remember that raising a number to a negative power, like a^{-x}, is equivalent to $\dfrac{1}{a^x}$. For example, $2^{-3} = \dfrac{1}{2^3} = \dfrac{1}{8}$.

The graph for this equation has a shape like this:

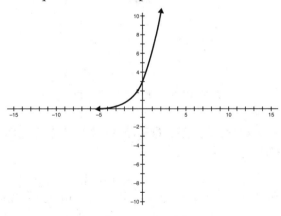

- For exponential decay, like $y = 3 \cdot \left(\frac{1}{2}\right)^x$, the graph has a shape like this:

Exponential equations can also be graphed on the graphing calculator.

8.4 REAL-WORLD SCENARIOS INVOLVING EXPONENTIAL EQUATIONS

Many real-world scenarios can be modeled with exponential equations. The population of a country over time is generally an example of exponential growth. The temperature of food over time after being put into a freezer is an example of exponential decay.

Practice Exercises

1. If $x = 2$ and $y = 3^x$, solve for y.
   - (1) 8
   - (2) 9
   - (3) 10
   - (4) 11

2. If $x = 3$ and $y = 2 \cdot 3^x$, solve for y.
   - (1) 216
   - (2) 27
   - (3) 54
   - (4) 18

3. Which ordered pair is in the solution set of $y = 5 \cdot 2^x$?
   - (1) $(0, 0)$
   - (2) $(2, 25)$
   - (3) $(3, 40)$
   - (4) $(3, 100)$

4. Which is the graph of $y = 2^x$?

(1)   (3)

(2)   (4)

5. Below is the graph of which equation?

(1) $y = \left(\dfrac{1}{3}\right)^x$ (3) $y = 4^x$

(2) $y = 10^x$ (4) $y = 3^x$

6. In what interval is the graph of $y = 1.5^x$ increasing?

(1) Always   (3) When $x \geq 0$

(2) Never   (4) When $x \leq 0$

7. Below is the graph of $y = b^x$. What is true about the value of b?

(1) b must be greater than 1

(2) b must be less than 1

(3) b must be less than 0

(4) b must be less than -1

8. What type of equation has a graph like the one below?

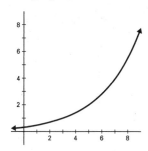

(1) Linear   (3) Quadratic

(2) Exponential   (4) None of the above

9. The population of a country can be modeled with the equation $P = 250 \cdot 1.07^t$, where P is the population in millions and t is the number of years since 2010. According to this model, rounded to the nearest ten million, what will the population of this country be in 2019?

(1) 450,000,000

(2) 460,000,000

(3) 470,000,000

(4) 480,000,000

10. A cup of tea that is 200 degrees is put into a room that is 80 degrees. The temperature of the tea can be calculated with the formula $t = 200 \cdot 0.9^m + 80$, where m is the number of minutes since the tea was put into the room. What will the temperature of the tea be after 15 minutes rounded to the nearest degree?

(1) 116 degrees

(2) 118 degrees

(3) 120 degrees

(4) 121 degrees

Solutions

1. Substitute 2 for x and the equation becomes $y = 3^2 = 9$.

The correct choice is **(2)**.

2. Substitute 3 for x and the equation becomes $y = 2 \cdot 3^3 = 2 \cdot 27 = 54$.

The correct choice is **(3)**.

3. Substitute each ordered pair into the equation to see which makes it true. For the ordered pair $(3, 40)$, $5 \cdot 2^3 = 5 \cdot 8 = 40$.

The correct choice is **(3)**.

4. The graph for $y = 2^x$ must pass through the points $(0, 1)$, $(1, 2)$, and $(2, 4)$. It also passes through $(-1, 1/2)$ and $(-2, 1/4)$.

The correct choice is **(3)**.

5. Since this graph passes through $(0, 1)$, $(1, 3)$, and $(2, 9)$, it is the equation $y = 3^x$.

The correct choice is **(4)**.

6. An exponential graph with a base greater than 1 is always increasing.

The correct choice is **(1)**.

7. When an exponential graph is decreasing, the base is between 0 and 1.

The correct choice is **(2)**.

8. This graph has the standard shape of an exponential curve based on an equation $y = b^x$ with $b > 1$. An exponential curve like this starts off relatively flat for x values close to zero and then increases rapidly.

   The correct choice is (**2**).

9. Since 2019 is 9 years past 2010, substitute 9 for t to get $P = 250 \cdot 1.07^9 = 459.6$. In millions, this is approximately 460,000,000.

   The correct choice is (**2**).

10. Substitute 15 for m to get $t = 200 \cdot 0.9^{15} + 80 = 121.1782$ or 121.2, which rounds to 121.

    The correct choice is (**4**).

9. CREATING AND INTERPRETING EQUATIONS

9.1 CREATING AND INTERPRETING LINEAR EQUATIONS

Many real-world situations can be modeled with a linear equation in the form $y = mx + b$. The m and the b have to be replaced with the appropriate values from the situation. In general, the b value is the starting value and the m value is the amount the total changes each time the x variable increases.

- If a carnival costs $10 admission and $3 for each ride, the 10 and the 3 can be used in a linear equation. Since the 10 is the starting amount, it would take the place of the b in the equation. Since 3 is the amount the total increases by for each new ride, it would take the place of m in the equation.

    The equation could be written $y = 3x + 10$, where y is the total and x is the number of rides. Instead of x and y, the total could be represented by the variable T, whereas the number of rides could be represented by the variable R to form the equation $T = 3R + 10$.

    When given an equation modeling a real-world situation, it is also possible to interpret what the values for m and b represent. The b represents the starting value, and the m represents the amount that the total changes for each increase in x.

- If the equation for the cost of a pizza with N toppings is $C = 2N + 12$, you could be asked to interpret what the 2 and the 12 represent. Since the 2 is in the place of the m in $y = mx + b$, it represents the cost of each topping. Since the 12 is in the place of the b in $y = mx + b$, it represents the cost of the pizza before any toppings are added.

9.2 CREATING AND INTERPRETING EXPONENTIAL EQUATIONS

An **exponential equation** is a good model for many real-world situations including population growth, compound interest, and liquid cooling. Exponential equations have the form $y = a \cdot (1 + r)^x$, where the a represents the starting value and the r represents the growth rate.

- If the population of a town is 10,000 people and the annual growth rate is 7%, then the equation that relates total population, P, to the number of years that have passed, T, is $P = 10{,}000 \cdot 1.07^T$. Since 10,000 is the starting value, it takes the place of the a and since the growth rate is .07, it takes the place of the r in the equation $y = a \cdot (1 + r)^x$.

- If an exponential equation that models a real-world situation is given, it is possible to interpret what the numbers in the equation represent.

- If, in another town, the equation relating their population to the number of years that have passed is $P = 20{,}000 \cdot 1.09^T$, you could be asked to interpret what the numbers 20,000 and .09 represent. In this case, the 20,000 represents the starting population and the .09 represents the growth rate.

- When a ball is dropped, each bounce is 80% as high as the previous bounce. If the ball is dropped from a window 50 feet above the ground, the equation that relates the height of the bounce (H) to the number of bounces (B) will be $H = 50 \cdot 0.80^B$. Since the starting height is 50, it takes the place of the a, and since 0.80 can be expressed as $(1 - 0.20)$, the r value is -0.20, which is the growth rate. A -0.20 growth rate can also be called a *decay* rate of 0.20.

Practice Exercises

1. It costs $10 to go to the movies and $3 for each bag of popcorn. Which equation relates the total cost (C) to the number of bags of popcorn purchased (P)?

   (1) $C = 3P + 10$

   (2) $C = 10P + 3$

   (3) $P = 3C + 10$

   (4) $P = 10C + 3$

2. A tablet computer costs $400 and $2 for each app. Which equation relates the total cost (C) to the number of apps purchased (A)?

   (1) $A = 2 + 400C$

   (2) $A = 400 + 2C$

   (3) $C = 2 + 400A$

   (4) $C = 400 + 2A$

3. A cable TV plan costs $80 a month plus $10 extra for each premium channel. Which equation relates the monthly bill (B) to the number of premium channels ordered (C)?

   (1) $B = 80C + 10$

   (2) $B = 10C + 80$

   (3) $C = 80B + 10$

   (4) $C = 10B + 80$

4. Lydia wants to buy a DVD player and some DVDs. The equation that relates the total cost for the DVD player and N DVDs is $P = 20N + 200$. What does the number 200 in the equation represent?

   (1) The cost of the DVD player

   (2) The cost of each DVD

   (3) The cost of all N DVDs

   (4) The total cost of the DVD player and all N DVDs

5. Amelia buys an empty sticker album and some sticker sheets. The equation that relates the total cost for the empty sticker album and N sticker sheets is $P = 0.75N + 3.00$. What does the number 0.75 in the equation represent?

   (1) The cost of the empty sticker album

   (2) The cost of each sticker sheet

   (3) The cost of all N sticker sheets

   (4) The total cost of the empty sticker album and all N sticker sheets

6. There were 900 birds in a forest. Each year the bird population increases by 12%. Which equation relates the bird population (P) to the number of years that have passed?

   (1) $P = 900(1.12)^t$ (3) $P = 900(0.88)^t$

   (2) $P = 900(0.12)^t$ (4) $t = 900(1.12)^P$

7. A bouncing ball is dropped from 20 feet high. After each bounce, the height of the next bounce is 65% as high as the last bounce. Which equation relates the height of the bounce (H) to the number of bounces that have happened (N)?

   (1) $H = 20(0.35)^N$ (3) $H = 20(0.65)^N$

   (2) $H = 20(1.65)^N$ (4) $N = 20(0.65)^H$

8. The population (P) of a town after t years can be modeled with the equation $P = 20,000(1.07)^t$. What does the 20,000 represent?

   (1) The growth rate

   (2) The percent increase each year

   (3) The population after t years

   (4) The starting population of the town

9. After Allie takes some medicine, each hour the number of milligrams of medicine (M) remaining in her body after t minutes can be modeled with the equation $M = 200(1 - 0.27)^t$. Which number represents the decay rate?

(1) 0.27

(3) 0.73

(2) 200

(4) 146

10. Mason puts money into a bank that offers interest compounded annually. The formula relating the amount of money in the bank (A) to the number of years it has been in the bank (t) is $A = 800(1.2)^t$. What is the interest rate the bank offers?

(1) 1.2%

(3) 20%

(2) 2%

(4) 120%

Solutions

1. P bags of popcorn cost \$$3P$. The entrance price is \$10 so the total price is $3P + 10$.

   The correct choice is (**1**).

2. A apps cost \$$2A$. The computer costs \$400 so the total price is $2A + 400$ or $400 + 2A$.

   The correct choice is (**4**).

3. C premium channels cost \$$10C$. The monthly cost is \$80 so the total price is $10C + 80$.

   The correct choice is (**2**).

4. The 20 is the cost for each DVD and the 200 is the price of the DVD player. In general, the constant represents the fixed cost.

   The correct choice is (**1**).

5. The 3 is the part that does not depend on the value of N. The $0.75N$ increases by 0.75 each time N increases by 1 so the 0.75 represents the cost of each sticker sheet.

   The correct choice is (**2**).

6. In an exponential equation of the form $P = a \cdot (1 + r)^t$, the r value is the percent increase (or decrease if r is negative), and the a is the initial value. When $r = 0.12$ and $a = 900$, the equation is $P = 900(1.12)^t$.

   The correct choice is (**1**).

7. The height of each bounce is 0.65 multiplied by the height of the previous bounce. The initial height is 20 feet. After one bounce, the second is $20 \cdot 0.65$. After two bounces it has a height of $20 \cdot 0.65 \cdot 0.65$. In general, after N bounces, the height will be $H = 20 \cdot 0.65^N$.

   The correct choice is (3).

8. In an equation of the form $P = a \cdot (1 + r)^t$, the r is the percent increase each year and the a is the initial value at $t = 0$. The 20,000 then represents the starting population of the town.

   The correct choice is (4).

9. In an equation of the form $M = a \cdot (1 - r)^t$, the r is the decay rate. Since the $(1 - 0.27)$ is being raised to the t power in this equation, the decay rate is 0.27.

   The correct choice is (1).

10. The interest rate is the percent increase each year. The percent increase in an equation of the form $A = P \cdot (1 + r)^t$ is the r variable. Since $1.2 = 1 + 0.2$, the r value is 0.2, which is 20%.

    The correct choice is (3).

10. FUNCTIONS

10.1 DIFFERENT REPRESENTATIONS OF FUNCTIONS

A **function** is like a machine that takes a number as an input and outputs a number. Functions are often named with lowercase letters like f and g.

- If a function f takes the number 2 as an input and outputs the number 7, we say $f(2) = 7$. The number in the parentheses is the number that is input into the function. The number after the equals sign is the number that is output from the function.

A function can be represented in several different ways. Here are the most common ways:

1. As a list of ordered pairs
- If the function f is defined as $f = \{(1, 4), (2, 7), (3, 10), (4, 13), (5, 16)\}$, the numbers in the parentheses represent an input value and an output value for the function. In this example, the point $(1,4)$ in the definition means that if 1 is put into the function, 4 is output from the function, or $f(1) = 4$. Likewise, $f(2) = 7, f(3) = 10, f(4) = 13$, and $f(5) = 16$.

2. As an equation
- A function can be defined by an equation that allows you to calculate the output value for a given input value. An example is the definition $f(x) = 3x + 1$. With this definition, it is possible to calculate the output value for any input value. For example, to calculate $f(10)$, substitute the number 10 into the equation $f(x) = 3x + 1$ to become $f(10) = 3(10) + 1 = 30 + 1 = 31$ so $f(10) = 31$.

3. As a graph
- If the function is defined as a graph, determine the value of $f(2)$ by finding a point on the graph that has an x-coordinate of 2. The y-coordinate of that point is the value of $f(2)$. Since the point with x-coordinate of 2 is $(2, 7)$ the value of $f(2) = 7$.

10.2 DOMAIN AND RANGE OF FUNCTIONS

The possible input values of a function are called the *domain* of the function. The possible output values are called the *range* of the function.

For the function $f = \{(1, 4), (2, 7), (3, 10), (4, 13), (5, 16)\}$, the domain is the set of input values $\{1, 2, 3, 4, 5\}$. The range is the set of output values $\{4, 7, 10, 13, 16\}$.

- When a function is described as a graph, the domain is the set of x-coordinates of all the points on the graph, and the range is the set of y-coordinates of all the points on the graph.

For this graph, the domain is $\{1, 2, 3, 4\}$, and the range is $\{3, 5, 8\}$.

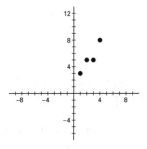

For this graph, the domain is $2 \leq x \leq 5$, and the range is $3 \leq y \leq 7$.

In a real-world situation, the domain is often a special subset of the real numbers. If the function has as its input the number of cars a salesman sells in a month, the domain would be the set of nonnegative integers $\{0, 1, 2, ...\}$ since you cannot sell fractions of a car or negative cars.

10.3 GRAPHING FUNCTIONS

The graph of the function $f(x) = x^2$ is the same as the graph $y = x^2$. All the methods for graphing by hand or with the graphing calculator described earlier can be used for functions.

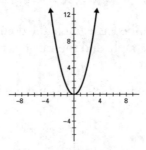

10.4 HOW TO TELL WHEN A GRAPH OR A LIST CANNOT DEFINE A FUNCTION

In a function, each time a number from the domain is put into the function, the same value is output from the function. So if $f(2) = 7$, the number 7 will always be output from the function whenever 2 is put into the function.

- When a function is defined as a list of ordered pairs, then there will never be two ordered pairs with the same x-coordinate, but different y-coordinates.

$f = \{(1, 4), (2, 7), (2, 10), (3, 13)\}$ is not the definition of a function since $f(2)$ can be 7 or 10.

$g = \{(1, 4), (2, 7), (3, 7), (4, 13)\}$ is the definition of a function. The fact that $g(2) = 7$ and $g(3) = 7$ does not contradict the definition of a function. As long as there are no repeats of x-coordinates, there can be repeats of y-coordinates.

A graph will not be the graph of a function if there are two points that have the same x-coordinate, but different y-coordinates. Graphs that cannot be functions fail the **vertical line test**, which means that at least one vertical line will pass through two or more points. All the points it passes through will have the same x-coordinates but different y-coordinates.

These graphs cannot be graphs of functions since they fail the vertical line test at at least one location.

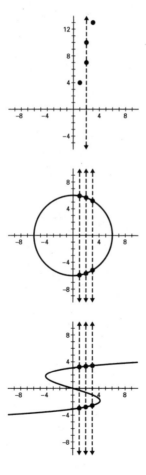

These graphs can be graphs of functions since they pass the vertical line test at all locations.

10.5 GRAPHING TRANSFORMED FUNCTIONS

- If the graph of $y = f(x)$ is already known, the graph of $y = f(x + a)$, $y = f(x - a)$, $y = f(x) + a$, and $y = f(x) - a$ can be easily graphed by knowing the four basic transformations.

- If the graph of $y = f(x)$ looks like this:

then the graph of

1. $f(x) + a$ will be the graph of $f(x)$ with every point shifted *up* by a units. The graph of $f(x) + 2$ looks like this:

2. $f(x) - a$ will be the graph of $f(x)$ with every point shifted *down* by a units. The graph of $f(x) - 2$ looks like this:

3. $f(x + a)$ will be the graph of $f(x)$ with every point shifted *left* by a units. The graph of $f(x + 2)$ looks like this:

4. $f(x - a)$ will be the graph of $f(x)$ with every point shifted *right* by a units. The graph of $f(x - 2)$ looks like this:

Practice Exercises

1. If a function f is defined as $f = \{(1, 2), (2, 3), (3, 1), (4, 4)\}$, what is $f(2)$?

    (1)   1   (3)   3

    (2)   2   (4)   4

2. Which of the following *cannot* be the definition of a function?

    (1) $f = \{(1, 5), (2, 7), (2, 8), (4, 9)\}$

    (2) $f = \{(1, 2), (2, 2), (3, 2), (4, 2)\}$

    (3) $f = \{(0, 0), (1, 1), (-1, 1), (2, 4), (-2, 4)\}$

    (4) $f = \{(6, 1)\}$

3. What is the domain of the function defined as $f = \{(1, 4), (3, 7), (4, 8), (5, 8)\}$?

    (1) $\{4, 7, 8\}$ (3) $\{1, 3, 4, 5\}$

    (2) $\{1, 3, 4, 5, 7, 8\}$ (4) $\{4\}$

4. Below is the graph of $y = f(x)$. What is the value of $f(3)$?

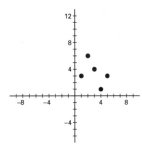

    (1)   1   (3)   3

    (2)   2   (4)   4

5. Which is the graph of a function?

(1)

(3)

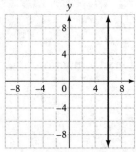

(2)

(4)

6. If $g(x) = -x^2 + 7x + 1$ what is $g(2)$?

(1) 11   (3) 27

(2) 19   (4) 35

7. If below is the graph of $y = f(x)$, which is the graph of $y = f(x) - 5$?

(1)

(3)

(2)

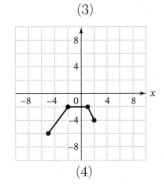

(4)

8. If $f(x) = 5x - 2$, what is $f(x + 1)$?

(1) $5x - 2$ (3) $5x + 1$

(2) $5x - 3$ (4) $5x + 3$

9. Below is the graph of $f(x)$ on the left and $g(x)$ on the right. Which is equivalent to $g(x)$?

(1) $f(x) + 4$ (3) $f(x + 4)$

(2) $f(x) - 4$ (4) $f(x - 4)$

10. If the graph of $y = f(x)$ is a parabola with the vertex at $(5, 1)$, what is the vertex of the graph of the parabola $y = f(x - 2)$?

(1) $(5, 3)$ (3) $(7, 1)$

(2) $(5, -1)$ (4) $(3, 1)$

Solutions

1. Since the ordered pair $(2, 3)$ is in the set, $f(2) = 3$.

   The correct choice is **(3)**.

2. A set of ordered pairs cannot be the definition of a function if there are two ordered pairs with the same x-coordinate, but different y-coordinates. In choice 1, there are two ordered pairs with x-coordinates of 2, $(2, 7)$ and $(2, 8)$, so that cannot be the definition of a function.

   The correct choice is **(1)**.

3. The domain is the set of x-coordinates. In this example, the x-coordinates are 1, 3, 4, and 5.

   The correct choice is **(3)**.

4. There is a point at $(3, 4)$ so $f(3) = 4$.

   The correct choice is **(4)**.

5. Choices 1, 2, and 4 all fail the vertical line test since at least one vertical line can pass through more than one point on them. Choice 3 can be the graph of a function since there is no possible vertical line that would pass through more than one point on it.

   The correct choice is **(3)**.

6. Substitute 2 for x to get $g(2) = -2^2 + 7 \cdot 2 + 1 = -4 + 14 + 1 = 11$.

   The correct choice is **(1)**. Note that $-2^2 = -4$ and not 4. If there were parentheses around -2, as in $(2)^2$, then it would equal 4.

7. The graph of $y = f(x) = -5$ is the same as the graph of $y = f(x)$ with each point shifted 5 units down.

   The correct choice is **(4)**.

8. Substitute $x + 1$ for x to get $f(x + 1) = 5(x + 1) - 2 = 5x + 5 - 2 = 5x + 3$.

   The correct choice is **(4)**.

9. The graph of $g(x)$ is the same as the graph of $y = f(x)$ with each point shifted 4 units to the right. $g(x)$ must be equivalent to $f(x - 4)$.

   The correct choice is **(4)**.

10. The graph of $f(x - 2)$ is the same as the graph of $f(x)$ with each point shifted two units to the right. If the vertex of the parabola for $f(x)$ is at $(5, 1)$, the vertex of the parabola for $f(x - 2)$ will be two units to the right of $(5, 1)$, which is at $(7, 1)$.

    The correct choice is **(3)**.

11. SEQUENCES

11.1 TYPES OF SEQUENCES

A **sequence** is a series of numbers that can be predicted by some kind of pattern. Two of the most common types of sequences are **arithmetic sequences** and **geometric sequences**.

- The sequence 2, 5, 8, 11, 14,... is an example of an arithmetic sequence since each term after the first can be obtained by adding the same number, 3, to the previous term.

- The sequence 2, 6, 18, 54, 162,... is an example of a geometric sequence since each term after the first can be obtained by multiplying the same number, 2, by the previous term.

- There are two notations that are used to describe the terms of a sequence. If the sequence is 2, 5, 8, 11, 14,... the first term can be described as either $a_1 = 2$ or $a(1) = 2$. The second notation is similar to function notation.

11.2 DESCRIBING A SEQUENCE WITH A DIRECT FORMULA

The terms of the sequence 2, 5, 8, 11,... can be described by the formula $a_n = 2 + 3(n - 1)$ or $a(n) = 2 + 3(n - 1)$. If you substitute $n = 1$ into the formula, it becomes $a_1 = 2 + 3(1 - 1) = 2 + 3(0) = 2 + 0 = 2$.

- For any arithmetic sequence, the direct formula for the nth term is $a_n = a_1 + d(n - 1)$ where a_1 is the first term of the sequence and d is the common difference between two consecutive terms. For the sequence 2, 5, 8, 11,..., $a_1 = 2$ and $d = 3$.

- For any geometric sequence, the direct formula for the nth term is $a_n = a_1 \cdot r^{n-1}$, where a_1 is the first term of the sequence and r is the common ratio between two consecutive terms. For the sequence 2, 6, 18, 54,..., $a_1 = 2$ and $r = 3$ since if you divide any term by the previous term you get 3.

11.3 DESCRIBING A SEQUENCE WITH A RECURSIVE FORMULA

Another way to describe the nth term of a sequence is to relate it to the previous term. The term before the a_n term is called the a_{n-1} term. A *recursive formula* for a sequence first defines the exact value of one or more of the terms and then describes how to obtain new terms from the previous ones.

- For the sequence 2, 5, 8, 11, 14, ... the recursive formula is

$$a_1 = 2$$
$$a_n = 3 + a_{n-1} \text{ for } n > 1$$

From the set of two equations, the entire sequence can be calculated. For example, when $n = 2$, $a_2 = a_1 + 3 = 2 + 3 = 5$. From this, the value of a_3 can be calculated, and this pattern can continue to obtain other n values. The recursive formulas are not convenient for large values of n, however.

- For the sequence 2, 6, 18, 54, 162, ..., the recursive formula is

$$a_1 = 2$$
$$a_n = 3 \cdot a_{n-1} \text{ for } n > 1$$

Practice Exercises

1. What type of sequence is $3, 7, 11, 15, \ldots$?
   - (1) Increasing arithmetic
   - (2) Decreasing arithmetic
   - (3) Increasing geometric
   - (4) Decreasing geometric

2. What type of sequence is $4, 8, 16, 32, \ldots$?
   - (1) Increasing arithmetic
   - (2) Decreasing arithmetic
   - (3) Increasing geometric
   - (4) Decreasing geometric

3. What is the next number in the sequence $20, 10, 5, \frac{5}{2}$?
   - (1) $\frac{5}{4}$
   - (2) $\frac{5}{8}$
   - (3) $\frac{5}{16}$
   - (4) $\frac{5}{32}$

4. Find the value of a_2 in the sequence defined by
$$a_1 = 4$$
$$a_n = 3 + a_{n-1} \text{ for } n > 1$$
   - (1) 3
   - (2) 5
   - (3) 7
   - (4) 12

5. The sequence $5, 11, 17, 23, \ldots$ can be generated by which definition?

(1) $a_1 = 5$
     $a_n = 6a_{n-1}$ for $n > 1$

(3) $a_1 = 5$
     $a_n = -6 + a_{n-1}$ for $n > 1$

(2) $a_1 = 5$
     $a_n = \dfrac{11}{5} a_{n-1}$ for $n > 1$

(4) $a_1 = 5$
     $a_n = 6 + a_{n-1}$ for $n > 1$

6. The sequence $3, 15, 75, 375, \ldots$ can be generated by which definition?

(1) $a_1 = 3$
     $a_n = 5a_{n-1}$ for $n > 1$

(3) $a_1 = 3$
     $a_n = \dfrac{1}{5} a_{n-1}$ for $n > 1$

(2) $a_1 = 3$
     $a_n = 12 + a_{n-1}$ for $n > 1$

(4) $a_1 = 3$
     $a_n = -12 + a_{n-1}$ for $n > 1$

7. A ball is dropped from a window 50 feet above the ground. Each bounce is $\dfrac{4}{5}$ the height of the previous bounce. Which definition would generate the height of the bounces?

(1) $a_1 = 50$
     $a_n = -10 + a_{n-1}$ for $n > 1$

(3) $a_1 = 50$
     $a_n = \dfrac{4}{5} a_{n-1}$ for $n > 1$

(2) $a_1 = 50$
     $a_n = 10 + a_{n-1}$ for $n > 1$

(4) $a_1 = 50$
     $a_n = \dfrac{5}{4} a_{n-1}$ for $n > 1$

8. What is the fifth term of the sequence generated by the definition $a_n = 10 - 4(n - 1)$?

(1) 18

(3) 2

(2) 12

(4) −6

9. What definition would produce the sequence 4, 13, 22, 31, ... ?

(1) $a_n = 4 + 9n$

(3) $a_n = 9 + 4n$

(2) $a_n = 4 + 9(n - 1)$

(4) $a_n = 9 + 4(n - 1)$

10. Which expression could be used to find the 20th term of the sequence 5, 9, 13, 17, 21, ... ?

(1) $5 + 4(20)$

(3) $4 + 5(20)$

(2) $5 + 4(19)$

(4) $4 + 5(19)$

Solutions

1. Since $7 = 3 + 4$ and $11 = 7 + 4$ and $15 = 11 + 4$, this is an increasing arithmetic sequence.

   The correct choice is **(1)**.

2. Since $8 = 2 \cdot 4$ and $16 = 2 \cdot 8$ and $32 = 2 \cdot 16$, this is an increasing geometric sequence.

   The correct choice is **(3)**.

3. Each term is equal to $\frac{1}{2}$ multiplied by the previous term. So the next term is $\frac{1}{2} \cdot \frac{5}{2} = \frac{5}{4}$.

   The correct choice is **(1)**.

4. $a_2 = 3 + a_{2-1} = 3 + a_1 = 3 + 4 = 7$.

   The correct choice is **(3)**.

5. The first term is a_1, so $a_1 = 5$. Each term is equal to 6 plus the previous term, so $a_n = 6 + a_{n-1}$.

   The correct choice is **(4)**.

6. The first term is a_1, so $a_1 = 3$. Each term is equal to 5 multiplied by the previous term, so $a_n = 5 \cdot a_{n-1}$.

   The correct choice is **(1)**.

7. The initial height is 50, so $a_1 = 50$. Each bounce is $4/5$ the height of the previous bounce so $a_n = 4/5 \cdot a_{n-1}$.

The correct choice is **(3)**.

8. Substitute 5 for n to get $a_5 = 10 - 4(5 - 1) = 10 - 4(4) = 10 - 16 = -6$.

The correct choice is **(4)**.

9. Substitute 1 for n in each equation. Only the second equation gets $a_1 = 4$. Also, an arithmetic sequence has the form $a_n = a_1 + d(n - 1)$, where d is the common difference and a_1 is the first term. Since $a_1 = 4$ and $d = 9$, the equation is $a_n = 4 + 9(n - 1)$.

The correct choice is **(2)**.

10. An arithmetic sequence with first term a_1 and common difference of d has the equation $a_n = a_1 + (n - 1)d$. a_1 is 5 in this example, and $d = 4$. Therefore, the equation is $a_n = 5 + 4(n - 1)$. Substitute 20 for n to get a $20 = 5 + 4(19)$.

The correct choice is **(2)**.

12. REGRESSION CURVES

12.1 THE LINE OF BEST FIT

A **line of best fit** is a line that comes as close as possible to a set of points on a graph. In the scatterplot below, there is no line that could pass through all ten points. Of all the possible lines, though, there is one that is a better fit than the others and this is the line of best fit.

The equation for the line of best fit can be determined quickly on a graphing calculator.

Below is a scatter plot and the chart on which it was based.

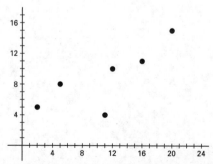

x	y
2	5
5	8
11	4
12	10
16	11
20	15

Instructions for the TI-84:

Press [STAT] and [1], and enter the x values into L1 and the y values into L2.

```
L1      L2      L3    2
2       5       ------
5       8
11      4
12      10
16      11
20      15
        ------
L2(7) =
```

Press [STAT], [right], [4], and [ENTER] for the equation of the line of best fit.

```
LinReg
y=ax+b
a=.4776785714
b=3.578869048
```

Instructions for the TI-Nspire:

From the home screen select the Add Lists and Spreadsheet icon. In the header row, label column A with an x and column B with a y. Enter the x values into cells A1 to A6. Enter the y values into cells B1 to B6.

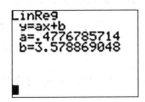

Press [menu], [4], [1], and [3] and select x in the X List field and y in the Y List field.

Press the OK button.

A x	B y	C	D
			=LinRegM
2	5	Title	Linear Re..
5	8	RegEqn	m*x+b
11	4	m	0.477679
12	10	b	3.57887
16	11	r²	0.617042

D1 = "Linear Regression (mx+b)"

The equation for the line of best fit is $y = 0.477679x + 3.57887$.

12.2 THE CORRELATION COEFFICIENT

The **correlation coefficient**, denoted by the variable r, is a number between -1 and 1. When a line of best fit has a positive slope and passes exactly through each of the points, the correlation coefficient is 1. When a line of best fit has a negative slope and passes exactly through each of the points, the correlation coefficient is -1. All correlation coefficients are between -1 and 1.

- The line of best fit for the scatter plot below is close to 1 because it has a positive slope and comes close to the points on the scatter plot. In this case $r = 0.9$

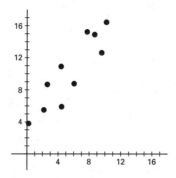

- The line of best fit for the following scatter plot is close to -1 because it has a negative slope and comes close to the points on the scatter plot, but not as close as the line in the previous example did to the points on that graph. In this case the $r = -0.8$

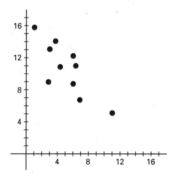

The graphing calculator can display the correlation coefficient. On the TI-84 it will be displayed along with the line of best fit only if the "diagnostics" are turned on. Press [2ND] and [0] to access the catalog and scroll to the DiagnosticOn command to do this. On the TI-Nspire it will display the correlation coefficient along with the equation of the line of best fit.

```
CATALOG          ▣
 DependAuto
 det(
 DiagnosticOff
▶DiagnosticOn
 dim(
 Disp
 DispGraph
```

```
LinReg
y=ax+b
a=.4776785714
b=3.578869048
r²=.6170415349
r=.7855199138
```

The correlation coefficient for the example from Section 12.1 was $r = 0.78552$.

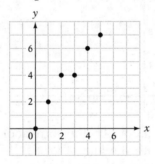

12.3 RESIDUAL PLOTS

When a line is a good fit for a scatterplot, the points in the plot are close to the points on the line. One way to measure this is to calculate the *residuals* and to examine them on a *residual plot*.

Here is a scatterplot with five points $(1, 2)$, $(2, 4)$, $(3, 4)$, $(4, 6)$, $(5, 7)$.

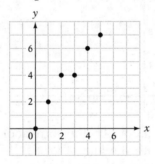

The line of best fit for this scatterplot is $y = 1.2x + 1$.

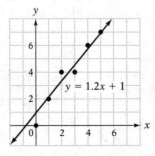

Only one of the points is on the line. For the others, two are above the line and two are below. Residuals measure how far above or below the line the points are. The information can be collected on a chart.

x	y	$1.2x + 1$	Residuals (y column) $-$ ($1.2x + 1$ column)
1	2	2.2	-0.2
2	4	3.4	0.6
3	4	4.6	-0.6
4	6	5.8	0.2
5	7	7	0

The two points above the line have positive residuals. The two points below the line have negative residuals.

A residual plot is a scatter plot with the x-coordinates from the x column in the chart and the y-coordinates from the residuals column in the chart.

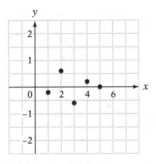

MATH FACTS

When the original scatterplot resembles a line, the residual plot will be a random scattering of points with no apparent pattern. When the original scatterplot resembles some other kind of curve, the residual plot will not be a bunch of random points but will look like a line or some kind of curve like a line or a U shape.

Example 1

Construct a residual plot for the scatter plot based on the following chart. What does this residual plot suggest about the original scatterplot?

Solution: The line of best fit is $y = 1.453x - 1.987$.

x	y	$1.453x - 1.987$	Residuals $(y \text{ column}) - (1.453x - 1.987)$
1	0.42	-0.534	0.954
2	0.62	0.919	-0.299
3	1.22	2.372	-1.152
4	3.21	3.825	-0.615
5	6.39	5.278	1.112

The residual plot looks like this:

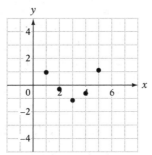

Because this has a U shape and is not just a random scattering of points, it suggests that the original scatter plot did not resemble a line.

12.4 PARABOLAS AND EXPONENTIALS OF BEST FIT

Even though every scatter plot has a line of best fit, sometimes even the best possible line isn't a very good fit.

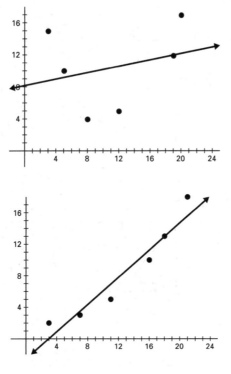

If the points in a scatter plot look more like a parabola or an exponential curve, the most appropriate curve to model it may not be a line. For parabolas, there is a parabola of best fit, and for exponential curves, there is an exponential curve of best fit. Both can be calculated on the graphing calculator. Select quadratic regression for the parabola of best fit and exponential regression for the exponential curve of best fit.

For the TI-84:

Enter the data into L1 and L2. Then press [STAT], [right], and [5].

  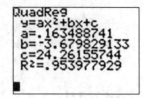

For the TI-Nspire:

To find the parabola of best fit, after entering the data into column A and column B, press [menu], [4], [1], and [6].

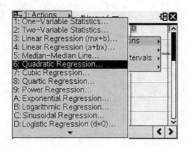

To find the exponential curve of best fit on the TI-84, enter the data into L1 and L2 and press [STAT], [right], and [0].

  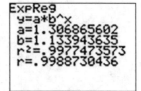

To find the exponential curve of best fit on the TI-Nspire, enter the data into column A and column B, press [menu], [4], [1], and [A].

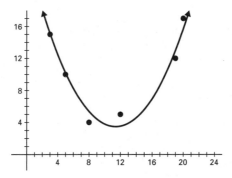

The parabola of best fit for the first scatter plot is $y = 0.1634989x^2 - 3.67983x + 24.2616$.

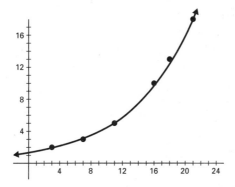

The exponential curve of best fit for the second scatterplot is $y = 1.30687 \cdot 1.13394^x$.

Practice Exercises

1. Calculate the equation for the line of best fit for the following set of data in $y = mx + b$ form. Round m and b to the nearest tenth.

x	y
1	3
2	5
3	4
4	6
5	8

(1) $y = 1.1x + 1.9$
(2) $y = 1.4x + 1.7$
(3) $y = 1.7x + 1.4$
(4) $y = 1.9x + 1.1$

2. Calculate the equation for the line of best fit for the following set of data in $y = mx + b$ form. Round m and b to the nearest tenth.

x	y
10	33
20	20
30	10
40	14
50	6

(1) $y = 34.6x - 0.6$

(2) $y = -0.6x + 34.6$

(3) $y = 31.8x - 0.7$

(4) $y = -0.7x + 31.8$

3. What is the equation for the line of best fit for the points on this scatter plot?

(1) $y = -0.5x + 7$

(2) $y = 0.7x - 0.5$

(3) $y = -0.4x + 0.9$

(4) $y = 0.9x - 0.4$

4. Of these four choices, which line appears to be the best fit for this scatter plot?

5. What is the correlation coefficient (r), rounded to the nearest hundredth, for the line of best fit for the data on the table below?

x	y
3	10
6	13
9	27
12	38
15	40

(1) 0.97

(2) 0.96

(3) 0.95

(4) 0.94

6. For which scatter plot is the correlation coefficient closest to 1?

(1)

(3)

(2)

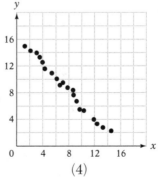

(4)

7. Of the four choices, which is closest to the correlation coefficient for this scatter plot?

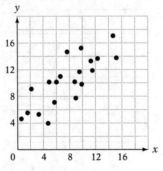

(1) 0.8   (3) 1
(2) −0.8   (4) −1

8. Find the equation of the parabola of best fit for the data on the table below.

x	y
1	2
2	2
3	3
4	4
5	6
6	8
7	10
8	14

(1) $y = 0.31x^2 - 67x + 3.53$
(2) $y = 0.24x^2 - 0.52x + 2.23$
(3) $y = 0.58x^2 - 0.41x + 4.12$
(4) $y = 0.73x^2 - 0.72x + 6.87$

9. Find the equation of the exponential curve of best fit for the data on the table below.

x	y
1	2
2	2
3	3
4	4
5	6
6	8
7	10
8	14

(1) $y = 1.28 \cdot 1.35^x$

(2) $y = 1.35 \cdot 1.28^x$

(3) $y = 1.47 \cdot 1.39^x$

(4) $y = 1.39 \cdot 1.47^x$

10. For which scatter plot would an exponential curve of best fit be most appropriate?

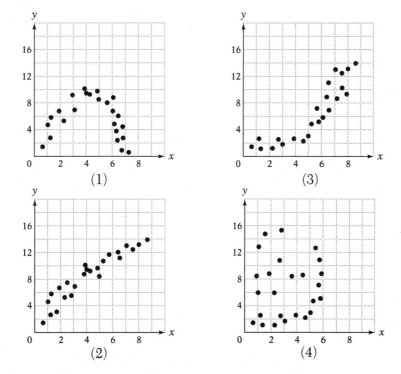

Solutions

1. Enter the data into the graphing calculator and do linear regression to get the equation $y = 1.1x + 1.9$.

   The correct choice is (**1**).

2. Enter the data into the graphing calculator and do linear regression to get the equation $y = -0.6x + 34.6$.

   The correct choice is (**2**).

3. Enter 2, 4, 6, 8, 10 for the x values and 1, 5, 3, 7, 9 for the y values into the graphing calculator and do linear regression. The equation is $y = 0.9x - 0.4$.

   The correct choice is (**4**).

4. Of the four choices, choice 3 seems to have the line closer to the points than the other three choices.

   The correct choice is (**3**).

5. Enter the data into the graphing calculator and do linear regression to get an r value of 0.97.

   The correct choice is (**1**).

6. To have an r value close to $+1$, the points must lie close to a line with a positive slope. Choices 1 and 2 both resemble lines with positive slopes. In choice 2 the points seem to fall closer to a straight line than choice 1.

   The correct choice is (**2**).

7. Since this scatter plot resembles a line with a positive slope, choices 2 and 4 can be eliminated because r must be positive. To have an r value of $+1$ the points would have to lie perfectly on a line, so choice 3 can be eliminated too.

The correct choice is (**1**).

8. Enter the data into the graphing calculator and do quadratic regression to get $y = 0.24x^2 - 0.52x + 2.23$.

The correct choice is (**2**).

9. Enter the data into the graphing calculator and do exponential regression to get $y = 1.28 \cdot 1.35^x$.

The correct choice is (**1**).

10. Of the four choices, choice 3 looks the most like an exponential curve. Choice 1 looks like a parabola. Choice 2 looks like a line. Choice 4 looks like a bunch of random points.

The correct choice is (**3**).

13. STATISTICS

13.1 MEAN, MEDIAN, AND MODE

In a set of numbers, the **mean**, also known as the **average**, is the sum of the numbers divided by how many numbers there are. For the set 70, 75, 78, 80, 82, the mean can be calculated as $\dfrac{70 + 75 + 78 + 80 + 82}{5} = \dfrac{385}{5} = 77$.

The **median** is the middle number (or average of the two middle numbers if there are an even amount of numbers) when the numbers are arranged from least to greatest. In the set 70, 75, 78, 80, 82, the median is the number 78. If the set were 70, 75, 78, 80, 82, 85, the median would be the average of 78 and 80, which is $\dfrac{78 + 80}{2} = 79$.

The **mode** is the number that appears most frequently. In the set 70, 75, 78, 80, 80, the mode is 80 since there are two 80s and only one of each of the other numbers.

13.2 FIRST QUARTILE AND THIRD QUARTILE

The median is greater than 50% of the numbers in the list. The number that is greater than just 25% of the numbers on the list is called the *first quartile*. The number that is greater than 75% of the numbers on the list is called the *third quartile*. To find the first quartile, find the median of all the numbers less than the median of the list. To find the third quartile, find the median of all the numbers greater than the median of the list. The interquartile range is the difference between the third quartile and the first quartile.

- For the list 40, 43, 45, 47, 48, 52, 57, 60, 61, 64, 68:
- The numbers less than the median 52 are 40, 43, 45, 47, 48. The median of these five numbers is 45, which is the first quartile.
- The numbers greater than the median 52 are 57, 60, 61, 64, 68. The median of these five numbers is 61, which is the third quartile. The interquartile range is $61 - 45 = 16$.

13.3 BOX PLOTS

A box plot is a picture that shows five different metrics, minimum, first quartile, median, third quartile, and maximum. To make a box plot, draw a segment connecting the maximum and the minimum. Then draw a rectangle with sides at the first quartile and third quartile. Finally, draw a line through the rectangle at the median.

- For the following numbers, the box plot looks like this:

    10, 10, 10, 12, 12, 12, 12, 17, 25, 25, 25, 25, 30, 30, 30

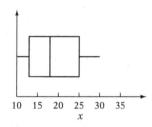

13.4 USING THE GRAPHING CALCULATOR TO DETERMINE MAXIMUM, MINIMUM, MEDIAN, FIRST QUARTILE, AND THIRD QUARTILE

For the TI-84:

The graphing calculator can calculate the five measures of central tendency. First enter all the numbers into L1 by pressing [STAT] and [1] for Edit.

To find the minimum, first quartile, median, third quartile, and maximum of the seven numbers 10, 4, 8, 12, 6, 16, 14, enter them into L1. Then press [STAT] and [1] for 1-Var Stats and press [ENTER].

The screen will display

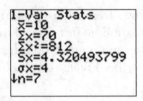

The $\bar{x} = 10$ is for the mean. The $n = 7$ means that there were seven elements in the list. For the minimum, first quartile, median, third quartile, and maximum, press the down arrow five times.

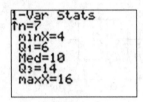

minX is for the minimum, Q1 is for the first quartile, Med is for the median, Q3 is for the third quartile, and maxX is for the maximum element.

For the TI-Nspire:

From the home screen, select the Add Lists & Spreadsheet icon. Name column A x and fill in cells A1 through A7 with the numbers 10, 4, 8, 12, 6, 16, 14.

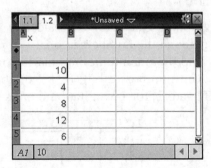

Press [4], [1], and [1] for One-Variable Statistics.

Press [OK] since there is just one list. Set the X1 List to x since that was what the column with the data was named in the spreadsheet.

In cells C2 through C13, the one-variable statistics will be displayed. The median is the $\bar{x}$-bar. n is for the number of numbers. MinX is the smallest number. Q1X is the first quartile. MedianX is the median. Q3X is the third quartile. MaxX is the largest number.

Practice Exercises

1. Find the mean of this set of numbers $\{4, 5, 8, 8, 8, 10, 10, 13, 15, 17, 23\}$.
   (1) 8   (3) 10
   (2) 9   (4) 11

2. Find the median of this set of numbers $\{4, 5, 8, 8, 8, 10, 10, 13, 15, 17, 23\}$.
   (1) 8   (3) 11
   (2) 10   (4) 15

3. Find the interquartile range of this set of numbers $\{4, 5, 8, 8, 8, 10, 10, 13, 15, 17, 23\}$.
   (1) 7   (3) 9
   (2) 8   (4) 10

4. For the first four days of a five-day vacation, the mean temperature was 80 degrees. What must the temperature be on the fifth day in order for the mean temperature to be 82 degrees?
   (1) 88   (3) 90
   (2) 89   (4) 91

5. For which data set is the median greater than the mean?
   (1) $\{4, 7, 10, 13, 16\}$ (3) $\{8, 9, 10, 11, 12\}$
   (2) $\{8, 9, 10, 18, 19\}$ (4) $\{1, 2, 10, 11, 12\}$

6. What is the mode of the data in this histogram?

(1) 23   (3) 15
(2) 24   (4) 16

7. What is the median of the data in this box plot?

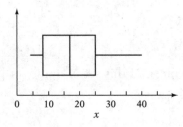

(1) 17   (3) 8
(2) 15.2   (4) 25

8. What is true about this data set: 1, 2, 10, 11, 11?

(1) The median is greater than the mean.
(2) The median is equal to the mean.
(3) The median is less than the mean.
(4) The median is greater than the mode.

9. For which data set is the interquartile range equal to 0?

    (1) 2, 3, 6, 6, 6, 7, 8

    (2) 2, 6, 6, 6, 6, 6, 8

    (3) 1, 2, 3, 6, 7, 8, 9

    (4) 2, 6, 6, 6, 6, 7, 8

10. In a set of seven numbers, the largest number is increased by 10. Which measure of central tendency must increase because of this?

    (1) Mean

    (2) Mode

    (3) First quartile

    (4) Interquartile range

Solutions

1. To get the mean, add the 11 values and divide by 11. $(4 + 5 + 8 + 8 + 8 + 10 + 10 + 13 + 15 + 17 + 23)/11 = 121/11 = 11$.

   The correct choice is (**4**).

2. The median is the middle number after the numbers have been arranged from least to greatest. These numbers are already arranged from least to greatest so the middle number is the sixth number, which is 10.

   The correct choice is (**2**).

3. The interquartile range is the difference between the first quartile and the third quartile. The median is the sixth number so the first quartile is the median of the first five numbers. The first five numbers are 4, 5, 8, 8, 8 with a median of 8. The third quartile is the median of the last five numbers. The last five numbers are 10, 13, 15, 17, 23, with a median of 15. The first quartile is 8. The third quartile is 15. The interquartile range is $15 - 8 = 7$.

   The correct choice is (**1**).

4. The sum of a set of numbers is the product of the average and the number of numbers. If the first four days had an average of 80, the sum of all four temperatures was 320. If the first five days need to have an average of 82, the sum of all five temperatures needs to be $82 \cdot 5 = 410$. The fifth temperature needs to be $410 - 320 = 90$.

   The correct choice is (**3**).

5. The median for all four data sets is 10. The mean of the numbers in choice 4 is $(1 + 2 + 10 + 11 + 12)/5 = 7.2$, which is less than the median. All the other choices have the mean greater than or equal to the median.

   The correct choice is (**4**).

6. The mode is the most frequent number. In a histogram the most frequent number is the tallest bar, which is 16 in this example.

The correct choice is (**4**).

7. The vertical line in the middle of the rectangle represents the median in a box plot. Since this is at 17, the median is 17.

The correct choice is (**1**).

8. The median of the data set is 10. The mode is 11. The mean is 7. The median is greater than the mean.

The correct choice is (**1**).

9. The interquartile range is zero if the first quartile is equal to the third quartile. This is the case for choice 2 where both the first quartile and the third quartile are 6.

The correct choice is (**2**).

10. When the largest number is increased by 10, the sum of all the numbers is increased by 10 so the mean will increase. The other measures will not be affected by increasing the largest number.

The correct choice is (**1**).

Glossary of Terms

Addition property of equality A property of algebra that states that when equal values are added to both sides of a true equation, the equation continues to be true. To solve the equation $x - 2 = 5$, add 2 to both sides of the equation by using the addition property of equality.

Arithmetic sequence A number sequence in which the difference between two consecutive terms is a constant. The sequence 2, 5, 8, 11, 14,...is an arithmetic sequence because the difference between consecutive terms is always 3.

Axis of symmetry An imaginary vertical line that passes through the vertex of a parabola. The equation for the axis of symmetry of a parabola is defined by $y = ax^2 + bx + c$ is $x = \frac{-b}{2a}$.

Base The number being raised to a power in an exponential expression. In the expression $2 \cdot 3^x$, the 3 is the base.

Binomial A polynomial with only two terms. $3x + 5$ is a binomial.

Box plot A graphical way to summarize data. The five numbers represented by the minimum, first quartile, median, third quartile, and maximum are graphed on a number line. A line segment connects the minimum to the first quartile. A rectangle is drawn around the first quartile and third quartile with a vertical line at the median. A line segment connects the third quartile to the maximum.

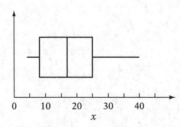

Closed form defined sequence A formula that defines the nth term of a sequence. The formula $a_n = 3 + 2(n - 1)$ is a closed form definition of a sequence. To get the 50th term of the sequence, substitute 50 for n in the definition.

Coefficient A number multiplied by a variable expression. In the expression $5x + 2$, 5 is the coefficient of x.

Common difference In an arithmetic sequence, the difference between consecutive terms. The common difference in the sequence 2, 5, 8, 11, 14, ... is 3.

Common ratio In a geometric sequence, the ratio between consecutive terms. The common ratio in the sequence 2, 6, 18, 54, 162, ... is 3 since $162/54 = 54/18 = 18/6 = 6/2 = 3$.

Commutative property of addition or multiplication The law from arithmetic that states that the order in which two numbers are added or multiplied does not matter. Because of the commutative property of addition, $5 + 2 = 2 + 5$.

Completing the square A method of solving a quadratic equation that involves turning one side of the equation into a perfect square trinomial.

Constant A number that does not have a variable part. In the expression $5x + 2$, 2 is a constant.

Correlation coefficient A number represented by r that measures how well a curve of best fit matches the points in a scatter plot. When the correlation coefficient is very close to 1 or to -1, the curve is a very good fit.

Degree of a polynomial The exponent on the highest power of a polynomial. In the polynomial $x^3 - 2x^2 + 5x - 2$, the degree is 3 since the highest power is a 3.

Difference of perfect squares Factoring a quadratic binomial in the form $x^2 - a^2$ into $(x - a)(x + a)$. For example, $x^2 - 9 = (x - 3)(x + 3)$.

Distributive property of multiplication over addition The rule that allows expressions of the form $a(b + c)$ to become $a \cdot b + a \cdot c$. For example, $2(3x + 5) = 6x + 10$.

Division property of equality A property of algebra that states that when both sides of a true equation are divided by the same non-zero number, the equation continues to be true. To solve the equation $2x = 8$, divide both sides of the equation by 2 using the division property of equality.

Domain The numbers that can be input into a function. When the function is defined as a set of ordered pairs or as a graph, the domain is the set of x-coordinates.

Dot plot A graphical way of representing a data set where each piece of data is represented with a dot.

Elimination method A way of solving a system of linear equations by combining the two equations in such a way as to eliminate one of the variables.

In the set of equations,

$$x + 2y = 12$$
$$3x - 2y = -4$$

the y variable is eliminated by adding the two equations together.

Equation Two mathematical expressions with an equals sign between them. $3x + 2 = 8$ is an equation.

Exponential equation An equation in which the variable is an exponent. $2 \cdot 3^x = 18$ is an exponential equation.

Exponential function A function in which the variable is an exponent. $f(x) = 2 \cdot 3^x$ is an exponential function.

Expression Numbers and variables that are combined with the operations from math.

Factoring a polynomial Finding two polynomials that can be multiplied to become another polynomial. The polynomial $x^2 + 5x + 6$ can be factored into $(x + 2)(x + 3)$.

Factors The polynomials that evenly divide into a polynomial. The factors of $x^2 + 5x + 6$ are $(x + 2)$ and $(x + 3)$.

First quartile The number in a data set that is bigger than just 25% of the numbers in the set.

FOIL A way of multiplying two binomials of the form $(a + b)(c + d)$ where (F)irst the a and c are multiplied, then the (O)uters a and d are multiplied, then the (I)nners b and c are multiplied, and finally the (L)asts b and d are multiplied. Then the four results are added together. The product of $(x + 2)$ and $(x + 3)$ is $x^2 + 3x + 2x + 6 = x^2 + 5x + 6$ by this method.

Function Something that takes numbers as inputs and outputs numbers. Functions are often labeled with the letters f or g. The notation $f(2) = 7$ means that when the number 2 is input into function f, it outputs the number 7.

Geometric sequence A number sequence in which the ratio between two consecutive terms (what you get when you divide one term by the term before it) is a constant. The sequence 2, 6, 18, 54, 162, ... is a geometric sequence because the ratio between consecutive terms is always 3.

Graph A visual way to describe the solution set to an equation. Each solution to the equation corresponds to an ordered pair that is graphed as a point on the coordinate plane. Each of the ordered pairs that satisfy an equation produce a point on the coordinate plane and the collection of all the points is the graph of the equation.

Greatest common factor The largest expression that divides evenly into two or more monomials. The greatest common factor of $6x^2$ and $8x^3$ is $2x^2$.

Growth rate In an exponential expression $a \cdot b^x$, the b is the growth rate. For example, in the equation $y = 500 \cdot 1.05^x$, the growth rate is 1.05.

Histogram A way of representing data with repeated values. Each value is represented by a bar whose height corresponds to the number of times that value is repeated. There are no spaces between the bars.

Increasing A function is increasing on an interval if making the input value larger also makes the y output larger. On a graph, an increasing function "goes up" from left to right.

Inequality Like an equation, but there is a $<$, $>$, $\leq$, or $\geq$ sign between the two expressions. $x + 2 > 5$ in a one-variable inequality. $y \leq 2x + 6$ is a two-variable inequality.

Interquartile range The difference between the number that is the third quartile of a data set and the number that is the first quartile of a data set.

Isolating a variable A variable is isolated when it is by itself on one side of an equation. In the equation $x + 2 = 5$, the x is not yet isolated. Subtracting 2 from both sides of the equation transforms the original equation into $x = 3$ with the x now isolated.

Like terms Terms that have the same variable part. They can be combined by adding or subtracting. $2x^2$ and $3x^2$ are like terms. $2x^2$ and $3x^3$ are not like terms.

Line of best fit A line that comes closest to the set of points in a scatter plot.

Linear equation An equation in which the greatest exponent is a 1. The equation $2x + 3 = 7$ is a linear equation.

Linear function A function in which the greatest exponent is a 1. The function $f(x) = 2x + 3$ is a linear function.

Mean The average of the numbers in a data set. Calculate the mean by adding all the numbers and dividing the total by the number of numbers in the set.

Median The middle number in a data set after it has been arranged from least to greatest. If there are an even number of numbers in the data set, the median is found by adding the two middle numbers and dividing by 2.

Mode The most frequent number in a data set.

Monomial A mathematical expression that has a coefficient and/or a variable part. $3x^2$ is a monomial. $3 + x^2$ is not a monomial.

Multiplication property of equality A property of algebra that states that when equal values are multiplied by both sides of a true equation, the equation continues to be true. To solve the equation $(1/2)x = 5$, multiply both sides of the equation by 2, using the multiplication property of equality.

Ordered pair Two numbers written in the form (x, y). An ordered pair can be a solution to a two-variable equation. For example, $(2, 5)$ is one solution to the equation $y = 2x + 1$. Ordered pairs can be graphed on the coordinate axes by locating the point with the x-coordinate equal to the x value and the y-coordinate equal to the y value.

Parabola A "U"-shaped curve that is the graph of the solution set of a quadratic equation.

Perfect square trinomial A quadratic trinomial of the form $x^2 + bx + (b/2)^2$, which can be factored into $(x + b/2)^2$. For example, $x^2 + 6x + 9 = (x + 3)^2$.

Piecewise function A function that has multiple rules for determining output values from input values, depending on what the input values are. If the function $f(x)$ is defined as

$$f(x) = 2x + 1 \quad \text{if } x < 0$$
$$x^2 \qquad\quad \text{if } x \geq 0$$

then $f(-3) = 2(-3) + 1 = -1$ and $f(5) = 5^2 = 25$.

Polynomial The sum of one or more monomials. Each term of the polynomial has the form ax^n. $3x^4 + 2x^3 - 3x^2 + 6x - 1$ is a polynomial.

Quadratic equation An equation in which the highest power on a variable is a 2. $x^2 + 5x + 6 = 0$ is a quadratic equation.

Quadratic formula The formula $x = \left(-b \pm \sqrt{\left(b^2 - 4ac\right)}\right)/2a$. This formula can be used to find the two solutions to the quadratic equation $ax^2 + bx + c = 0$.

Quadratic function A function in which the highest power on a variable is 2. $f(x) = x^2 + 5x + 6$ is a quadratic function.

Quadratic polynomial A polynomial in which the highest power on a variable is 2. $x^2 + 5x + 6$ is a quadratic polynomial.

Range The set of values that can be output from a function is the range of that function.

Recursively defined sequence A way to define a sequence in which the first term or terms of the sequence are given and a formula is given for calculating the next term based on the previous term or terms.

$$a_1 = 5$$
$$a_n = 3 + a_{n-1} \text{ for } n > 1$$

is a recursive definition for the sequence 5, 8, 11, 14, ...

Regression Finding a curve that best fits a scatterplot. Three types of regression are linear, quadratic, and exponential.

Residual plot A set of points that represent how far points on a graph deviate from a curve of best fit.

Roots The roots of an equation are the values that solve that equation. The roots of $x^2 + 5x + 6 = 0$ are -3 and -2.

Sequence A list of numbers that usually have some kind of pattern.

Slope The slope of a line is a measure of the line's steepness. The equation for the slope of a line that passes through the two points (x_1, y_1) and (x_2, y_2) is

$$m = \frac{y_2 - y_1}{x_2 - x_1}.$$

Slope-intercept form An equation in the form $y = mx + b$ where m and b are numbers in slope intercept form. $y = 2x - 1$ is in slope intercept form. When a two-variable equation is in slope intercept form, the graph of the equation has a y-intercept of $(0, b)$ and a slope of m.

Solution set The set of numbers or ordered pairs that satisfies an equation. The solution set of $x + 2 = 5$ is $\{3\}$. The solution set of $x + y = 10$ has an infinite number of ordered pairs in its solution set, including $(2, 8)$, $(3, 7)$, and $(4, 6)$.

Substitution method A method for solving a system of equations in which one variable is isolated in one of the equations, and the expression equal to that variable is substituted for it in the other equation.

Subtraction property of equality A property of algebra that states that when equal values are subtracted from both sides of a true equation, the equation continues to be true. To solve the equation $x + 2 = 5$, subtract 2 from both sides of the equation by using the subtraction property of equality.

System of equations Two or more equations with two or more unknowns to solve for. An example of a system of equations with a solution of $(8, 2)$ is

$$x + y = 10$$
$$x - y = 6$$

Third quartile The number in a data set that is greater than just 75% of the numbers in the set.

Translation A translation of a graph is when the points on it are each shifted the same amount in the same direction. Examples of translations are vertical translations, horizontal translations, and combinations of vertical and horizontal translations.

Trinomial A polynomial with three terms. The polynomial $x^2 + 5x + 6$ is a trinomial.

Variable A letter, often an x, y, or z, that represents a value in a mathematical expression. In an algebraic equation, the variable is often the unknown that needs to be solved for.

Vertex The turning point of a parabola is its vertex. If the parabola opens upward, the vertex is the minimum point. If the parabola opens downward, the vertex is the maximum point.

Vertical line test A way of testing to see if a graph can represent a function. If at least one vertical line can pass through at least two points on the graph, the graph fails the vertical line test and cannot represent a function. If there are no vertical lines that can pass through at least two points, then the graph can be the graph of a function.

x-intercept The location where a curve crosses the x-axis. The y-coordinate of the x-intercept is 0.

y-intercept The location where a curve crosses the y-axis. The x-coordinate of the y-intercept is 0. In slope-intercept form, $y = mx + b$, the y-intercept is located at $(0, b)$.

Zeros The zeros of a function f are the numbers that can be input into the function so that 0 is output from the function. For example, the function $f(x) = 2x - 6$ has the number 3 as its only zero since $f(3) = 2(3) - 6 = 6 - 6 = 0$.

Regents Examinations, Answers, and Self-Analysis Charts

Examination
June 2017
Algebra I

HIGH SCHOOL MATH REFERENCE SHEET

Conversions

1 inch = 2.54 centimeters

1 meter = 39.37 inches

1 mile = 5280 feet

1 mile = 1760 yards

1 mile = 1.609 kilometers

1 kilometer = 0.62 mile

1 pound = 16 ounces

1 pound = 0.454 kilogram

1 kilogram = 2.2 pounds

1 ton = 2000 pounds

1 cup = 8 fluid ounces

1 pint = 2 cups

1 quart = 2 pints

1 gallon = 4 quarts

1 gallon = 3.785 liters

1 liter = 0.264 gallon

1 liter = 1000 cubic centimeters

Formulas

Triangle	$A = \frac{1}{2}bh$
Parallelogram	$A = bh$
Circle	$A = \pi r^2$
Circle	$C = \pi d$ or $C = 2\pi r$

Formulas (continued)

General Prisms	$V = Bh$
Cylinder	$V = \pi r^2 h$
Sphere	$V = \frac{4}{3}\pi r^3$
Cone	$V = \frac{1}{3}\pi r^2 h$
Pyramid	$V = \frac{1}{3}Bh$
Pythagorean Theorem	$a^2 + b^2 = c^2$
Quadratic Formula	$x = \dfrac{-b \pm \sqrt{b^2 - 4ac}}{2a}$
Arithmetic Sequence	$a_n = a_1 + (n-1)d$
Geometric Sequence	$a_n = a_1 r^{n-1}$
Geometric Series	$S_n = \dfrac{a_1 - a_1 r^n}{1 - r}$ where $r \neq 1$
Radians	1 radian $= \dfrac{180}{\pi}$ degrees
Degrees	1 degree $= \dfrac{\pi}{180}$ radians
Exponential Growth/Decay	$A = A_0 e^{k(t-t_0)} + B_0$

PART I

Answer all 24 questions in this part. Each correct answer will receive 2 credits. No partial credit will be allowed. For each statement or question, write in the space provided the numeral preceding the word or expression that best completes the statement or answers the question.   **[48 credits]**

1 To keep track of his profits, the owner of a carnival booth decided to model his ticket sales on a graph. He found that his profits only declined when he sold between 10 and 40 tickets. Which graph could represent his profits?

(1)

(3)

(2)

(4)

1 _____

2 The formula for the surface area of a right rectangular prism is $A = 2lw + 2hw + 2lh$, where l, w, and h represent the length, width, and height, respectively. Which term of this formula is *not* dependent on the height?

(1) A (3) $2hw$

(2) $2lw$ (4) $2lh$ 2 _____

3 Which graph represents $y = \sqrt{x} - 2$?

(1)

(3)

(2)

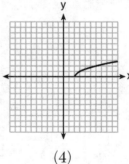

(4)   3 _____

4 A student plotted the data from a sleep study as shown in the
graph below.

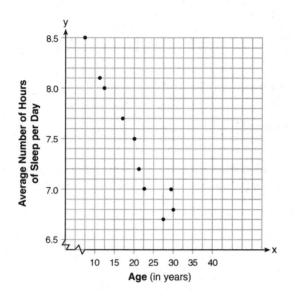

The student used the equation of the line

$$y = -0.09x + 9.24$$

to model the data. What does the rate of change represent in terms
of these data?

(1) The average number of hours of sleep per day increases
0.09 hour per year of age.

(2) The average number of hours of sleep per day decreases
0.09 hour per year of age.

(3) The average number of hours of sleep per day increases
9.24 hours per year of age.

(4) The average number of hours of sleep per day decreases
9.24 hours per year of age.   4 _____

5 Lynn, Jude, and Anne were given the function $f(x) = -2x^2 + 32$, and they were asked to find $f(3)$. Lynn's answer was 14, Jude's answer was 4, and Anne's answer was ±4. Who is correct?

(1) Lynn, only   (3) Anne, only

(2) Jude, only   (4) Both Lynn and Jude   5 _____

6 Which expression is equivalent to $16x^4 - 64$?

(1) $(4x^2 - 8)^2$ (3) $(4x^2 + 8)(4x^2 - 8)$

(2) $(8x^2 - 32)^2$ (4) $(8x^2 + 32)(8x^2 - 32)$ 6 _____

7 Vinny collects population data, $P(h)$, about a specific strain of bacteria over time in hours, h, as shown in the graph below.

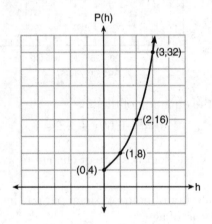

Which equation represents the graph of $P(h)$?

(1) $P(h) = 4(2)^h$

(2) $P(h) = \dfrac{46}{5}h + \dfrac{6}{5}$

(3) $P(h) = 3h^2 + 0.2h + 4.2$

(4) $P(h) = \dfrac{2}{3}h^3 - h^2 + 3h + 4$ 7 _____

8 What is the solution to the system of equations below?

$$y = 2x + 8$$
$$3(-2x + y) = 12$$

(1) no solution   (3) $(-1, 6)$

(2) infinite solutions   (4) $\left(\dfrac{1}{2}, 9\right)$ 8 _____

9 A mapping is shown in the diagram below.

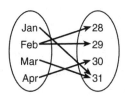

This mapping is

(1) a function, because Feb has two outputs, 28 and 29
(2) a function, because two inputs, Jan and Mar, result in the output 31
(3) not a function, because Feb has two outputs, 28 and 29
(4) not a function, because two inputs, Jan and Mar, result in the output 31   9 _____

10 Which polynomial function has zeros at $-3, 0$, and 4?

(1) $f(x) = (x + 3)(x^2 + 4)$
(2) $f(x) = (x^2 - 3)(x - 4)$
(3) $f(x) = x(x + 3)(x - 4)$
(4) $f(x) = x(x - 3)(x + 4)$ 10 _____

11 Jordan works for a landscape company during his summer vacation. He is paid \$12 per hour for mowing lawns and \$14 per hour for planting gardens. He can work a maximum of 40 hours per week, and would like to earn at least \$250 this week. If m represents the number of hours mowing lawns and g represents the number of hours planting gardens, which system of inequalities could be used to represent the given conditions?

(1) $m + g \leq 40$
$12m + 14g \geq 250$

(3) $m + g \leq 40$
$12m + 14g \leq 250$

(2) $m + g \geq 40$
$12m + 14g \leq 250$

(4) $m + g \geq 40$
$12m + 14g \geq 250$

11 _____

12 Anne invested \$1000 in an account with a 1.3% annual interest rate. She made no deposits or withdrawals on the account for 2 years. If interest was compounded annually, which equation represents the balance in the account after the 2 years?

(1) $A = 1000(1 - 0.013)^2$

(3) $A = 1000(1 - 1.3)^2$

(2) $A = 1000(1 + 0.013)^2$

(4) $A = 1000(1 + 1.3)^2$

12 _____

13 Which value would be a solution for x in the inequality $47 - 4x < 7$?

(1) -13

(3) 10

(2) -10

(4) 11

13 _____

14 Bella recorded data and used her graphing calculator to find the equation for the line of best fit. She then used the correlation coefficient to determine the strength of the linear fit.

Which correlation coefficient represents the strongest linear relationship?

(1) 0.9

(3) -0.3

(2) 0.5

(4) -0.8

14 _____

15 The heights, in inches, of 12 students are listed below.

61, 67, 72, 62, 65, 59, 60, 79, 60, 61, 64, 63

Which statement best describes the spread of these data?

(1) The set of data is evenly spread.

(2) The median of the data is 59.5.

(3) The set of data is skewed because 59 is the only value below 60.

(4) 79 is an outlier, which would affect the standard deviation of these data.

15 _____

16 The graph of a quadratic function is shown below.

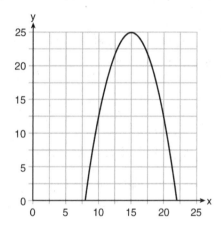

An equation that represents the function could be

(1) $q(x) = \frac{1}{2}(x + 15)^2 - 25$

(2) $q(x) = -\frac{1}{2}(x + 15)^2 - 25$

(3) $q(x) = \frac{1}{2}(x - 15)^2 + 25$

(4) $q(x) = -\frac{1}{2}(x - 15)^2 + 25$

16 _____

17 Which statement is true about the quadratic functions $g(x)$, shown in the table below, and $f(x) = (x - 3)^2 + 2$?

x	g(x)
0	4
1	−1
2	−4
3	−5
4	−4
5	−1
6	4

(1) They have the same vertex.

(2) They have the same zeros.

(3) They have the same axis of symmetry.

(4) They intersect at two points.   17 _____

18 Given the function $f(n)$ defined by the following:

$$f(1) = 2$$
$$f(n) = -5f(n - 1) + 2$$

Which set could represent the range of the function?

(1) $\{2, 4, 6, 8, \ldots\}$

(2) $\{2, -8, 42, -208, \ldots\}$

(3) $\{-8, -42, -208, 1042, \ldots\}$

(4) $\{-10, 50, -250, 1250, \ldots\}$ 18 _____

19 An equation is given below.

$$4(x - 7) = 0.3(x + 2) + 2.11$$

The solution to the equation is

(1) 8.3   (3) 3

(2) 8.7   (4) −3   19 _____

20 A construction worker needs to move 120 ft^3 of dirt by using a wheelbarrow. One wheelbarrow load holds 8 ft^3 of dirt and each load takes him 10 minutes to complete. One correct way to figure out the number of hours he would need to complete this job is

(1) $\dfrac{120\,\text{ft}^3}{1} \cdot \dfrac{10\,\text{min}}{1\,\text{load}} \cdot \dfrac{60\,\text{min}}{1\,\text{hr}} \cdot \dfrac{1\,\text{load}}{8\,\text{ft}^3}$

(2) $\dfrac{120\,\text{ft}^3}{1} \cdot \dfrac{60\,\text{min}}{1\,\text{hr}} \cdot \dfrac{8\,\text{ft}^3}{10\,\text{min}} \cdot \dfrac{1}{1\,\text{load}}$

(3) $\dfrac{120\,\text{ft}^3}{1} \cdot \dfrac{1\,\text{load}}{10\,\text{min}} \cdot \dfrac{8\,\text{ft}^3}{1\,\text{load}} \cdot \dfrac{1\,\text{hr}}{60\,\text{min}}$

(4) $\dfrac{120\,\text{ft}^3}{1} \cdot \dfrac{1\,\text{load}}{8\,\text{ft}^3} \cdot \dfrac{10\,\text{min}}{1\,\text{load}} \cdot \dfrac{1\,\text{hr}}{60\,\text{min}}$

20 _____

21 One characteristic of all linear functions is that they change by

(1) equal factors over equal intervals

(2) unequal factors over equal intervals

(3) equal differences over equal intervals

(4) unequal differences over equal intervals

21 _____

22 What are the solutions to the equation $x^2 - 8x = 10$?

(1) $4 \pm \sqrt{10}$ (3) $-4 \pm \sqrt{10}$

(2) $4 \pm \sqrt{26}$ (4) $-4 \pm \sqrt{26}$

22 _____

23 The formula for blood flow rate is given by $F = \dfrac{p_1 - p_2}{r}$, where F is the flow rate, p_1 the initial pressure, p_2 the final pressure, and r the resistance created by blood vessel size. Which formula can *not* be derived from the given formula?

(1) $p_1 = Fr + p_2$ (3) $r = F(p_2 - p_1)$

(2) $p_2 = p_1 - Fr$ (4) $r = \dfrac{p_1 - p_2}{F}$ 23 _____

24 Morgan throws a ball up into the air. The height of the ball above the ground, in feet, is modeled by the function $h(t) = -16t^2 + 24t$, where t represents the time, in seconds, since the ball was thrown. What is the appropriate domain for this situation?

(1) $0 \le t \le 1.5$ (3) $0 \le h(t) \le 1.5$

(2) $0 \le t \le 9$ (4) $0 \le h(t) \le 9$ 24 _____

PART II

Answer all 8 questions in this part. Each correct answer will receive 2 credits. Clearly indicate the necessary steps, including appropriate formula substitutions, diagrams, graphs, charts, etc. For all questions in this part, a correct numerical answer with no work shown will receive only 1 credit.   [16 credits]

25   Express in simplest form: $(3x^2 + 4x - 8) - (-2x^2 + 4x + 2)$

26 Graph the function $f(x) = -x^2 - 6x$ on the set of axes below.

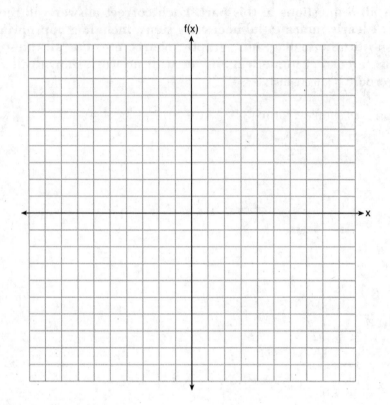

State the coordinates of the vertex of the graph.

27 State whether $7 - \sqrt{2}$ is rational or irrational. Explain your answer.

28 The value, $v(t)$, of a car depreciates according to the function $v(t) = P(.85)^t$, where P is the purchase price of the car and t is the time, in years, since the car was purchased. State the percent that the value of the car *decreases* by each year. Justify your answer.

29 A survey of 100 students was taken. It was found that 60 students watched sports, and 34 of these students did not like pop music. Of the students who did *not* watch sports, 70% liked pop music. Complete the two-way frequency table.

	Watch Sports	Don't Watch Sports	Total
Like Pop			
Don't Like Pop			
Total			

30 Graph the inequality $y + 4 < -2(x - 4)$ on the set of axes below.

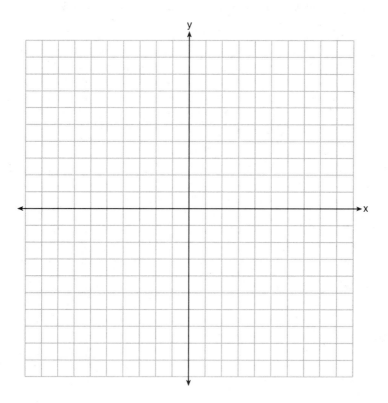

31 If $f(x) = x^2$ and $g(x) = x$, determine the value(s) of x that satisfy the equation $f(x) = g(x)$.

32 Describe the effect that each transformation below has on the function $f(x) = |x|$, where $a > 0$.

$g(x) = |x - a|$

$h(x) = |x| - a$

PART III

Answer all 4 questions in this part. Each correct answer will receive 4 credits. Clearly indicate the necessary steps, including appropriate formula substitutions, diagrams, graphs, charts, etc. For all questions in this part, a correct numerical answer with no work shown will receive only 1 credit.   [16 credits]

33 The function $r(x)$ is defined by the expression $x^2 + 3x - 18$. Use factoring to determine the zeros of $r(x)$.

Explain what the zeros represent on the graph of $r(x)$.

34 The graph below models Craig's trip to visit his friend in another state. In the course of his travels, he encountered both highway and city driving.

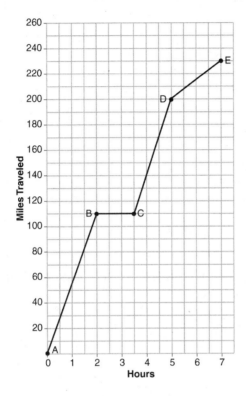

Based on the graph, during which interval did Craig most likely drive in the city? Explain your reasoning.

Question 34 is continued on the next page.

Question 34 continued

Explain what might have happened in the interval between B and C.

Determine Craig's average speed, to the *nearest tenth of a mile per hour*, for his entire trip.

35 Given:

$$g(x) = 2x^2 + 3x + 10$$
$$k(x) = 2x + 16$$

Solve the equation $g(x) = 2k(x)$ algebraically for x, to the *nearest tenth*.

Explain why you chose the method you used to solve this quadratic equation.

36   Michael has $10 in his savings account. Option 1 will add $100 to his account each week. Option 2 will double the amount in his account at the end of each week.

Write a function in terms of x to model each option of saving.

Michael wants to have at least $700 in his account at the end of 7 weeks to buy a mountain bike. Determine which option(s) will enable him to reach his goal. Justify your answer.

PART IV

Answer the question in this part. A correct answer will receive 6 credits. Clearly indicate the necessary steps, including appropriate formula substitutions, diagrams, graphs, charts, etc. A correct numerical answer with no work shown will receive only 1 credit.   [6 credits]

37 Central High School had five members on their swim team in 2010. Over the next several years, the team increased by an average of 10 members per year. The same school had 35 members in their chorus in 2010. The chorus saw an increase of 5 members per year.

Write a system of equations to model this situation, where x represents the number of years since 2010.

Question 37 is continued on the next page.

Question 37 continued

Graph this system of equations on the set of axes below.

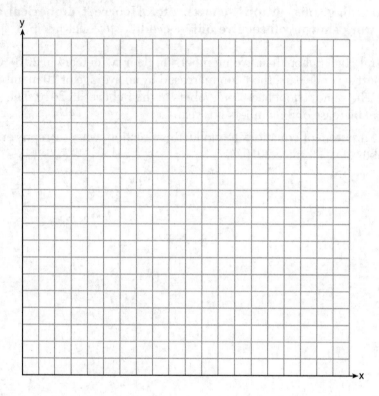

Explain in detail what each coordinate of the point of intersection of these equations means in the context of this problem.

Answers
June 2017
Algebra I

Answer Key

PART I

1. (3)	**5.** (1)	**9.** (3)	**13.** (4)	**17.** (3)	**21.** (3)
2. (2)	**6.** (3)	**10.** (3)	**14.** (1)	**18.** (2)	**22.** (2)
3. (4)	**7.** (1)	**11.** (1)	**15.** (4)	**19.** (1)	**23.** (3)
4. (2)	**8.** (1)	**12.** (2)	**16.** (4)	**20.** (4)	**24.** (1)

PART II

25. $5x^2 - 10$

26. Vertex: $(-3, 9)$

27. Irrational

28. 15%

29.

	Watch Sports	Don't Watch Sports	Total
Like Pop	26	28	54
Don't Like Pop	34	12	46
Total	60	40	100

30. See graph

31. $\{0, 1\}$

32. $g(x)$ is shifted a units to the right. $h(x)$ is shifted a units down.

PART III

33. The zeros are -6 and 3. They represent the x-intercepts of the graph.

34. Between D and E, the speed is 15 miles per hour, which is reasonable for city driving. Between B and C, he was stopped. The average speed for the entire trip is 32.9 miles per hour.

35. $\{3.6, -3.1\}$ The quadratic formula is simplest.

36. Option 1: $y = 100x + 10$. Option 2: $y = 10 \cdot 2^x$. Both are over \$700 after 7 weeks.

PART IV

37. Swim team: $y = 10x + 5$. Chorus: $y = 5x + 35$. The intersection point is $(6, 65)$. In 2016, they both have 65 members.

In **Parts II–IV**, you are required to show how you arrived at your answers. For sample methods of solutions, see the *Answers Explained* section.

Answers and Explanations

PART I

1. If the profits are declining only when he sells between 10 and 40 tickets, the graph should show a curve that is going down as it goes from left to right between 10 and 40.

   The graph in choice (1) is flat between 10 and 40. The graph in choice (2) is going up between 10 and 40. The graph in choice (4) is going up from 10 to 25 but then down from 25 to 40. Only the graph in choice (3) is going down for the entire interval between 10 and 40.

   The correct choice is (**3**).

2. A term is dependent on a variable if a change in that variable's value would also change the value of that entire term. When a term contains a variable, the term is dependent on that variable.

   Choices (3) and (4) have the variable h in them, so those can be eliminated. Even though choice (1) doesn't look like it has an h in it, it does. If you look at the definition of A, its equation has two hs in it. So A is also dependent on h. Choice (2) has no h in it or in the definition of any of the variables involved in that choice.

   The correct choice is (**2**).

3. Graphing the relation $y = \sqrt{x - 2}$ on the graphing calculator makes a graph that resembles choice (**4**).

   For the TI-84:

For the TI-Nspire:

This graph can also be made without the graphing calculator by knowing that the graph of the function $y = \sqrt{x}$ has the following shape.

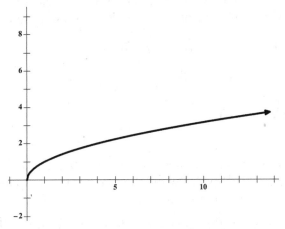

The graph of $y = \sqrt{x-2}$ is a transformation of the graph of $y = \sqrt{x}$ shifted 2 units to the right.

The correct choice is (**4**).

4. The rate of change is equivalent to the slope of the line. Since m represents the slope when the equation is of the form $y = mx + b$, the rate of change for this model is -0.09. For every 1 unit that a point on the line goes to the right, the line will go down by 0.09 units.

For this model, a change by 1 unit to the right corresponds to an increase in age by 1 year. A change of 0.09 units down corresponds to a 0.09 decrease in the average number of hours of sleep per day.

The correct choice is (**2**).

5. To calculate $f(3)$, substitute 3 for x in the equation:

$$f(3) = -2 \cdot 3^2 + 32$$
$$= -2 \cdot 9 + 32$$
$$= -18 + 32$$
$$= 14$$

Lynn got 14 for her answer. The other two people seem to have tried to solve the equation $0 = -2x^2 + 32$.

The correct choice is (**1**).

6. Since both $16x^4$ and 64 are perfect squares, this expression can be written as $(4x^2)^2 - 8^2$. This can be factored as the difference of perfect squares:

$$a^2 - b^2 = (a - b)(a + b)$$
$$\left(4x^2\right)^2 - 8^2 = \left(4x^2 - 8\right)\left(4x^2 + 8\right)$$

The correct choice is (**3**).

7. Each time the x-coordinate increases by 1 in this graph, the y-coordinate doubles. This is what happens when the graph is of an exponential equation, and only choice (1) is exponential.

Another way to solve this question is to calculate to see if $P(0) = 4$ and $P(1) = 8$ for any of the answer choices. Test choice (1):

$$P(0) = 4(2)^0 = 4 \cdot 1 = 4$$
$$P(1) = 4(2)^1 = 4 \cdot 2 = 8$$

The correct choice is (**1**).

8. Since the first equation has the y-variable isolated on one side of the equation, the simplest way to solve this system of equations is with the substitution method. Substitute $2x + 8$ for the y in the second equation:

$$3\left(-2x + (2x + 8)\right) = 12$$
$$3\left(-2x + 2x + 8\right) = 12$$
$$3(8) = 12$$
$$24 = 12$$

Since 24 can never equal 12, this system of equations has no solution.

The correct choice is (**1**).

9. In a function, each input value can be mapped to only one output value. When that happens, you can make statements like $f(\text{Jan}) = 31$ and $f(\text{Apr}) = 30$. In this mapping diagram though, Feb is mapped to two different numbers. So this mapping is not considered a function.

The correct choice is (**3**).

10. A zero of a polynomial is a number that when put into the function makes an output of zero. For -3 to be a zero, $(x + 3)$ must be a factor since $-3 + 3 = 0$. For 4 to be a zero, $(x - 4)$ must be a factor. For 0 to be a zero, x must be a factor.

So the function is $f(x) = x(x + 3)(x - 4)$.

The correct choice is (**3**).

11. Since Jordan can work at most 40 hours, the sum of his hours mowing and planting gardens must be less than or equal to 40. If he mows for m hours at \$12 per hour, the amount he makes mowing is $12m$. If he plants gardens for g hours at \$14 per hour, the amount he makes planting gardens is $14g$. Together he wants these amounts to be greater than or equal to 250.

So the system of inequalities is

$$m + \quad g \le 40$$
$$12m + 14g \ge 250$$

The correct choice is (**1**).

12. After the first deposit, Anne has $1000 in the account. After 1 year, she earns 1.3% interest. Her new balance can be calculated by multiplying $1000 by 0.013 to find the amount of interest. Then add the interest to the original $1000 to get $1000 + $13 = $1013. To find the amount Anne has after 2 years, multiply $1013 by 0.013 to find the amount of interest. Then add the interest to the $1013 to get $1013 + $13.169 = $1026.169. If you test each of the choices, only choice (2) gets this same answer.

A shorter way is to know that interest compounded annually can be calculated with the formula $A = P(1 + r)^n$, where P is the initial deposit, r is the percent interest as a decimal, and n is the number of years. In this example, $P = 1000$, $r = 0.013$, and $n = 2$.

The correct choice is (**2**).

13. Start solving the inequality the same way you would solve an equation that contained an equal sign instead of a $<$ sign:

$$47 - 4x < 7$$
$$-47 \quad\quad = -47$$
$$-4x < -40$$

When you divide both sides of an inequality by a negative, you must flip the direction of the inequality sign:

$$\frac{-4x}{-4} > \frac{-40}{-4}$$
$$x > 10$$

Of the choices, only the number 11 is greater than 10.

You could answer this question by just testing each of the answer choices to see which of them make the inequality true. Test choice (4):

$$47 - 4(11) < 7$$
$$47 - 44 < 7$$
$$3 < 7$$

This is true. Testing the other choices would result in an inequality that is false.

The correct choice is (**4**).

14. The closer the correlation coefficient is to either 1 or -1, the stronger the linear relationship is. Since 0.9 is just 0.1 away from 1 while -0.8 is 0.2 away from -1, 0.9 is the closest.

The correct choice is (**1**).

15. Arrange the numbers from smallest to largest:

$$59, 60, 60, 61, 61, 62, 63, 64, 65, 67, 72, 79$$

Analyze each of the answer choices.

- Choice (1): Evenly spaced numbers would be something like 59, 61, 63, 65, 67, where each number is the same amount more than the number before. That is not what is happening with this set of numbers.

- Choice (2): The median is the middle number. Since there are an even amount of numbers, there is no one middle number. For an even amount of numbers, the median is the average of the two middle numbers, in this case the 6th and 7th numbers:

$$\text{median} = \frac{62 + 63}{2} = 62.5$$

The median is 62.5, not 59.5. Note that 59.5 is what you would get if you did not arrange the numbers in order before taking the average of the 6th and 7th numbers.

- Choice (3): Most data would be considered skewed. In this case, the data are not skewed because 59 is the only number less than 60.

- Choice (4): The standard deviation is a measure of how close together the numbers are. If all the numbers were the same, the standard deviation would be zero. Without the 79, the standard deviation would be much lower since all the other numbers are pretty close to 65.

For the TI-84:

For the TI-Nspire:

The correct choice is (**4**).

16. The graph shows a parabola with a vertex at (15, 25) and that opens downward. The vertex form for the equation of a parabola is $y = a(x - h)^2 + k$, where h is the x-coordinate of the vertex, k is the y-coordinate of the vertex, a will be positive if the parabola opens upward, and a will be negative if the parabola opens downward. The graphs for choices (3) and (4) are parabolas with a vertex at (15, 25). Since choice (4) has a negative a-value, the parabola will open downward like in the question.

On the graphing calculator, the four choices can be graphed. Only choice (4) will resemble the graph in the question.

For the TI-84:

For the TI-Nspire:

The correct choice is (**4**).

17. The graphs of $f(x)$ and $g(x)$ look like this.

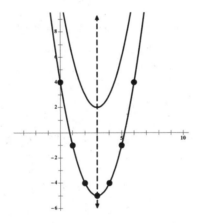

Analyze the answer choices:

- Choice (1): The vertex of $g(x)$ is $(3, -5)$, while the vertex of $f(x)$ is $(3, 2)$.
- Choice (2): Zeros can be seen on a graph as x-intercepts. Function $g(x)$ has two zeros. One is between 0 and 1, and the other is between 5 and 6. Function $f(x)$ does not have any zeros.
- Choice (3): The axis of symmetry of $g(x)$ is $x = 3$, and the axis of symmetry of $f(x)$ is also $x = 3$.
- Choice (4): The graphs do not intersect.

The correct choice is (**3**).

18. Since $f(1) = 2$, the first number in the sequence must be 2. Eliminate choices (3) and (4). To calculate the second number in the sequence, substitute 2 for n in the recursive part of the definition:

$$\begin{aligned} f(2) &= -5f(2-1) + 2 \\ &= -5f(1) + 2 \\ &= -5 \cdot 2 + 2 \\ &= -10 + 2 = -8 \\ &= -8 \end{aligned}$$

The correct choice is (**2**).

19. First use the distributive property:

$$4(x-7) = 0.3(x+2) + 2.11$$
$$4x - 28 = 0.3x + 0.6 + 2.11$$

Then combine like terms:

$$4x - 28 = 0.3x + 2.71$$

Complete the problem by getting the variables on one side of the equation and the constants on the other side of the equation:

$$\begin{aligned} 4x - 28 &= 0.3x + 2.71 \\ -0.3x &= -0.3x \\ 3.7x - 28 &= 2.71 \\ +28 &= +28 \\ \frac{3.7x}{3.7} &= \frac{30.71}{3.7} \\ x &= 8.3 \end{aligned}$$

The correct choice is (**1**).

20. The simplest way to solve this is to calculate the number of hours to complete the job. To calculate the number of wheelbarrow loads needed, divide 120 by 8 to get 15 wheelbarrows. To calculate the number of minutes needed to move the dirt, multiply the number of wheelbarrow loads by 10, which is $15 \times 10 = 150$ minutes. Finally, convert 150 minutes to hours by dividing 150 by 60 to get 2.5 hours.

    Test the four answer choices to see which one equals 2.5 hours.

    The correct choice is (**4**).

21. A linear function is something like $f(x) = 2x^1 - 3 = 2x - 3$. In other words, the exponent on the variable is 1. The graph of a linear function is a line.

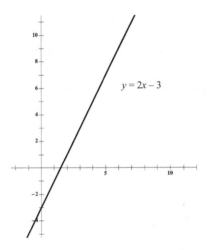

$y = 2x - 3$

Start from any point on the line, such as (2, 1), and pick an interval, such as 3. The point 3 to the right has the coordinates (5, 7). The interval in this case is 3. The difference in the y-coordinates is $7 - 1 = 6$. If you start from some different point, such as (4, 5), and choose the same interval, 3, the other point will be (7, 11). The difference in the y-coordinates is 6. The differences are the same.

The correct choice is (**3**).

22. There are three ways to solve this question.

One way is to complete the square by adding $\left(\dfrac{-8}{2}\right)^2 = (-4)^2 = 16$ to both sides of the equation:

$$x^2 - 8x = 10$$
$$+16 = +16$$
$$x^2 - 8x + 16 = 26$$
$$(x - 4)^2 = 26$$
$$\sqrt{(x - 4)^2} = \pm\sqrt{26}$$
$$x - 4 = \pm\sqrt{26}$$

Another way to answer this is to substitute each of the answers for x into the expression $x^2 - 8x$ to see which one makes the expression evaluate to 16.

For the TI-84:

For the TI-Nspire:

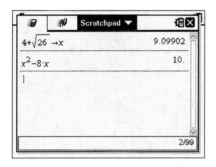

The third method is to use the quadratic formula:

$$x = \frac{-b \pm \sqrt{b^2 - 4ac}}{2a}$$

Remember to start by rearranging the given equation into the format $ax^2 + bx + c = 0$:

$$x^2 - 8x = 10$$

$$x^2 - 8x - 10 = 0$$

$$x = \frac{-(-8) \pm \sqrt{(-8)^2 - 4(1)(-10)}}{(2)(1)}$$

$$x = \frac{8 \pm \sqrt{64 + 40}}{2}$$

$$x = \frac{8 \pm \sqrt{104}}{2}$$

$$x = \frac{8 \pm \sqrt{4 \cdot 26}}{2}$$

$$x = \frac{8 \pm 2\sqrt{26}}{2}$$

$$x = 4 \pm \sqrt{26}$$

The correct choice is (**2**).

23. To isolate the other variables, start by multiplying both sides of the equation by r:

$$Fr = r \cdot \frac{p_1 - p_2}{r}$$

$$Fr = p_1 - p_2$$

The isolate r:

$$\frac{Fr}{F} = \frac{p_1 - p_2}{F}$$

This is choice (4).

Then isolate p_1:

$$Fr = p_1 - p_2$$

$$+p_2 = +p_2$$

$$Fr + p_2 = p_1$$

This is choice (1).

Then isolate p_2:

$$Fr = p_1 - p_2$$
$$+p_2 = +p_2$$
$$Fr + p_2 = p_1$$
$$-Fr = -Fr$$
$$p_2 = p_1 - Fr$$

This is choice (2).

Only choice (3) cannot be derived from the given formula.

The correct choice is (**3**).

24. Since the ball never goes underground, the appropriate domain is the values of t of which $h(t) \geq 0$. By graphing the function on a graphing calculator, it can be seen that the ball starts on the ground at $t = 0$ and lands when $t = 1.5$.

For the TI-84:

For the TI-Nspire:

This can also be solved using algebra by calculating the values of t that make $h(t) = 0$:

$$-16t^2 + 24t = 0$$
$$8t(-2t + 3) = 0$$
$$8t = 0 \quad \text{or} \quad -2t + 3 = 0$$
$$t = 0 \quad \text{or} \quad -2t = -3$$
$$t = \frac{3}{2}$$

If the ball is on the ground at $t = 0$ and at $t = \frac{3}{2} = 1.5$, then it is above the ground between those two values.

The correct choice is (**1**).

PART II

25. Eliminate the parentheses by distributing the negative sign through the second expression. Then combine like terms:

$$\left(3x^2 + 4x - 8\right) - \left(2x^2 + 4x + 2\right)$$
$$3x^2 + 4x - 8 - 2x^2 + 4x - 2$$
$$5x^2 - 10$$

26. The graph can be created with a table.

x	$f(x) = -x^2 - 6x$
-6	$-(-6)^2 - 6(-6) = -36 + 36 = 0$
-5	$-(-5)^2 - 6(-5) = -25 + 30 = 5$
-4	$-(-4)^2 - 6(-4) = -16 + 24 = 8$
-3	$-(-3)^2 - 6(-3) = -9 + 18 = 9$
-2	$-(-2)^2 - 6(-2) = -4 + 12 = 8$
-1	$-(-1)^2 - 6(-1) = -1 + 6 = 5$
0	$-(0)^2 - 6(0) = 0 + 0 = 0$

This table can also be created with the graphing calculator.

For the TI-84:

For the TI-Nspire:

 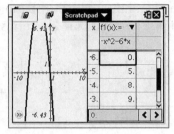

The graph looks like this:

The vertex is the highest point of the parabola (or the lowest if it is a "smiling" parabola). The coordinates of the vertex, according to the graph, are $(-3, 9)$.

You can also calculate the x-coordinate of the vertex by using the following formula:

$$x = \frac{-b}{2a}$$
$$= \frac{-(-6)}{2(-1)}$$
$$= \frac{6}{-2}$$
$$= -3$$

The y-coordinate of the vertex is what you get when you substitute the x-coordinate for x in the function:

$$y = f(-3)$$
$$= -(-3)^2 - 6(-3)$$
$$= -9 + 18$$
$$= 9$$

So the coordinates of the vertex are $(-3, 9)$.

27. When an irrational number is added to or subtracted from a rational number, the result will be irrational.

    If you calculate this difference on a calculator it becomes 5.585786..., which appears to be irrational since it does not terminate and does not seem to have a repeating pattern.

28. One way to solve this is to substitute a value for the initial price of the car, like $P = 10000$.

For $t = 0$:

$$v(0) = 10000(.85)^0$$
$$= 10000 \cdot 1$$
$$= 10000$$

One year later for $t = 1$:

$$v(1) = 10000(.85)^1$$
$$= 10000 \cdot .85$$
$$= 8500$$

Calculate the percent change from $t = 0$ to $t = 1$:

$$\% \text{ change} = \frac{\text{old price} - \text{new price}}{\text{old price}}$$
$$= \frac{10000 - 8500}{10000}$$
$$= \frac{15000}{1000}$$
$$= 0.15$$
$$= 15\%$$

A quicker way to solve this is to write the equation in the form $v(t) = P(1 + r)^t$, where r is the percent change. Since $.85 = (1 - .15)$, the percent change is 15%. Since r is negative, this is a percent decrease.

29. From the given information, three values from the chart can be quickly filled in.

	Watch Sports	**Don't Watch Sports**	**Total**
Like Pop			
Don't Like Pop	34		
Total	60		100

From these three values, the number of students who do not watch sports can be calculated by subtracting the number of students who watch sports from the total number of students:

$$100 - 60 = 40$$

	Watch Sports	Don't Watch Sports	Total
Like Pop			
Don't Like Pop	34		
Total	60	40	100

Since 70% of the students who do not watch sports do like pop music, calculate 70% of 40 and put that number into the table:

$$40 \cdot 0.7 = 28$$

	Watch Sports	Don't Watch Sports	Total
Like Pop		28	
Don't Like Pop	34		
Total	60	40	100

Now there is enough information to calculate the remaining four cells in the table. Each row and each column must add up to the total:

$$60 - 34 = 26$$
$$40 - 28 = 12$$
$$34 + 12 = 46$$
$$26 + 28 = 54$$

	Watch Sports	Don't Watch Sports	Total
Like Pop	26	28	54
Don't Like Pop	34	12	46
Total	60	40	100

30. First rewrite the inequality in slope-intercept form:

$$y + 4 < -2(x - 4)$$
$$y + 4 < -2x + 8$$
$$-4 = -4$$
$$y < -2x + 4$$

Graph the boundary line $y = -2x + 4$. Make it a dotted line because the inequality sign is $<$ and not $\leq$. It has a slope of -2 and a y-intercept of 4.

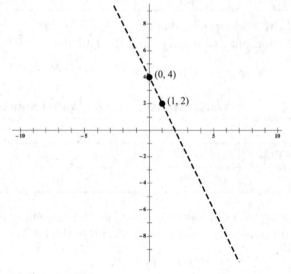

To decide which side of the line to shade, test to see if the point $(0, 0)$ makes the inequality true:

$$0 < -2(0) + 4$$
$$0 < 0 + 4$$
$$0 < 4$$

Since this is true, shade the side of the line containing (0, 0).

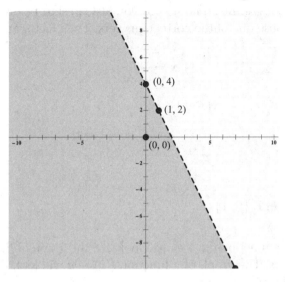

This can also be done on the graphing calculator.

For the TI-84:

For the TI-Nspire:

31. If $f(x) = g(x)$, then $x^2 = x$. To solve this equation, the first step is to subtract x from both sides of the equal sign. If you instead start by dividing both sides by x, you will lose one of the correct answers. Then factor and solve for x.

$$x^2 = x$$
$$-x = -x$$
$$x^2 - x = 0$$
$$x(x - 1) = 0$$

$$x = 0 \qquad \text{or} \qquad x - 1 = 0$$
$$+1 = +1$$
$$x = 1$$

The solution set is $\{0, 1\}$.

32. The graph of a function $f(x - a)$ will look like the graph of $f(x)$ shifted to the right by a units. The graph of a function $f(x) - a$ will look like the graph of $f(x)$ shifted down by a units.

By choosing a value for a, like $a = 2$, and graphing $f(x)$, $g(x)$, and $h(x)$ on a graphing calculator it is easier to see how the graphs of $g(x)$ and $h(x)$ compare to the graph of $f(x)$.

For the TI-84:

For the TI-Nspire:

So $g(x)$ is shifted a units to the right, and $h(x)$ is shifted a units down.

PART III

33. The zeros of a function are the input values that make the function equal to 0. To find zeros, solve the equation:

$$r(x) = 0$$
$$x^2 + 3x - 18 = 0$$

Factor a polynomial in the form $x^2 + bx + c$ into $(x + p)(x + q)$ by finding, if possible, two numbers that have a sum of b and a product of c. In this case since b is 3 and c is -18, the two numbers are 6 and -3 since $6 + -3 = 3$ and $6 \cdot -3 = 18$:

$$x^2 + 3x - 18 = 0$$
$$(x + 6)(x - 3) = 0$$

$$x + 6 = 0 \quad \text{or} \quad x - 3 = 0$$
$$-6 = -6 \qquad\qquad +3 = +3$$
$$x = -6 \qquad\qquad x = 3$$

On the graph of the function, the zeros are the x-intercepts. The x-intercepts of the graph of $r(x)$ are $(-6, 0)$ and $(3, 0)$. This can be seen on the graphing calculator.

For the TI-84:

For the TI-Nspire:

34. In a graph where the x-axis is time and the y-axis is distance, the slope of a line segment is the speed at which the object is moving. There are four line segments in this graph. Find the slope of each segment.

- Calculate the slope from A to B.

  From hours $= 0$ to hours $= 2$, the distance changed from 0 miles to 110 miles:

  $$\text{speed} = \text{slope}$$
  $$= \frac{110 - 0}{2 - 0}$$
  $$= \frac{110}{2}$$
  $$= 55 \text{ miles per hour}$$

- Calculate the slope from B to C.

  From hours $= 2$ to hours $= 3.5$, the distance stayed at 110 for the whole time:

  $$\text{speed} = \text{slope}$$
  $$= \frac{110 - 110}{3.5 - 2}$$
  $$= \frac{0}{1.5}$$
  $$= 0 \text{ miles per hour (stopped)}$$

- Calculate the slope from C to D.

  From hours $= 3.5$ to hours $= 5$, the distance changed from 110 miles to 200 miles:

  $$\text{speed} = \text{slope}$$
  $$= \frac{200 - 110}{5 - 3.5}$$
  $$= \frac{90}{1.5}$$
  $$= 60 \text{ miles per hour}$$

- Calculate the slope from D to E.

  From hours $= 5$ to hours $= 7$, the distance changed from 200 miles to 230 miles:

$$\begin{aligned} \text{speed} &= \text{slope} \\ &= \frac{230 - 200}{7 - 5} \\ &= \frac{30}{2} \\ &= 15 \text{ miles per hour} \end{aligned}$$

Craig most likely drove in the city between D and E because city driving is most likely to be 15 miles per hour.

In the interval between B and C, his speed was zero. So Craig did not move at all. He may have stopped to eat.

Since Craig travels a total of 230 miles in 7 hours, his average speed is $\frac{230}{7} \approx 32.9$ miles per hour.

35. Start by setting $g(x)$ and $2k(x)$ equal to each other:

$$\begin{aligned} g(x) &= 2k(x) \\ 2x^2 + 3x + 10 &= 2(2x + 16) \\ 2x^2 + 3x + 10 &= 4x + 32 \\ -4x \quad\quad &= -4x \\ 2x^2 - x + 10 &= 32 \\ -32 &= -32 \\ 2x^2 - x - 22 &= 0 \end{aligned}$$

Then use the quadratic formula:

$$x = \frac{-b \pm \sqrt{b^2 - 4ac}}{2a}$$

$$x = \frac{-(-1) \pm \sqrt{(-1)^2 - 4 \cdot 2 \cdot (-22)}}{2 \cdot 2}$$

$$x = \frac{1 \pm \sqrt{1 + 176}}{4} = \frac{1 \pm \sqrt{177}}{4}$$

$$x = \frac{1 + \sqrt{177}}{4} \approx 3.6 \quad \text{or} \quad x = \frac{1 - \sqrt{177}}{4} \approx -3.1$$

Of the three options: factoring, completing the square, and the quadratic formula, the quadratic formula was simplest for this example. The coefficient of 2 in front of the x^2 makes using the other two options complicated.

36. A function to model Option 1 is $f(x) = 10 + 100x$.

A function to model Option 2 is $g(x) = 10 \cdot 2^x$.

After 7 weeks with Option 1, he will have:

$$f(7) = 10 + 100 \cdot 7$$
$$= 10 + 700$$
$$= 710$$

After 7 weeks with Option 2, he will have:

$$g(7) = 10 \cdot 2^7$$
$$= 10 \cdot 128$$
$$= 1280$$

Both Option 1 and Option 2 will enable Michael to have more than $700. Either option will let him reach his goal.

PART IV

37. The equation for the swim team can be written as $y = 5 + 10x$ or $y = 10x + 5$.

The equation for the chorus can be written as $y = 35 + 5x$ or $y = 5x + 35$.

Before graphing, choose an appropriate scale. Since the point of intersection must be on the graph, you must find that point of intersection before deciding on the scale.

Using a chart is the simplest way to determine the approximate intersection point.

x	swim	chorus
0	5	35
1	15	40
2	25	45
3	35	50
4	45	55
5	55	60
6	65	65
7	75	70

Since the grid is 20 by 20 and the point (6, 65) must be on it, one possible scale is to have one box equal 0.5 units on the x-axis and 5 units on the y-axis.

The graph looks like this:

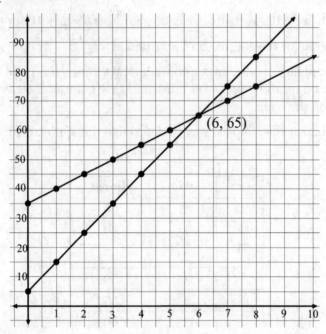

The point of intersection is (6, 65). It means that in the year 2016, both the swim team and the chorus had 65 members.

Topic	Question Numbers	Number of Points	Your Points	Your Percentage
1. Polynomials	5, 10, 25, 31, 33	$2 + 2 + 2 + 2 + 4 = 12$		
2. Properties of Algebra	2, 19, 23	$2 + 2 + 2 = 6$		
3. Functions	9, 17, 20, 21, 24, 32, 35	$2 + 2 + 2 + 2 + 2 + 2 + 4 = 16$		
4. Creating and Interpreting Equations	4, 28	$2 + 2 = 4$		
5. Inequalities	11, 13, 30	$2 + 2 + 2 = 6$		
6. Sequences and Series	18	2		
7. Systems of Equations	8, 37	$2 + 6 = 8$		
8. Quadratic Equations and Factoring	6, 22	$2 + 2 = 4$		
9. Regression	14	2		
10. Exponential Equations	7, 12, 36	$2 + 2 + 4 = 8$		
11. Graphing	1, 16, 26, 34	$2 + 2 + 2 + 4 = 10$		
12. Statistics	15, 29	$2 + 2 = 4$		
13. Number Properties	20, 27	$2 + 2 = 4$		

HOW TO CONVERT YOUR RAW SCORE TO YOUR ALGEBRA I REGENTS EXAMINATION SCORE

The accompanying conversion chart must be used to determine your final score on the June 2017 Regents Examination in Algebra I. To find your final exam score, locate in the column labeled "Raw Score" the total number of points you scored out of a possible 86 points. Since partial credit is allowed in Parts II, III, and IV of the test, you may need to approximate the credit you would receive for a solution that is not completely correct. Then locate in the adjacent column to the right the scale score that corresponds to your raw score. The scale score is your final Algebra I Regents Examination score.

Regents Examination in Algebra I—June 2017
Chart for Converting Total Test Raw Scores to Final
Examination Scores (Scaled Scores)

Raw Score	Scale Score	Performance Level	Raw Score	Scale Score	Performance Level	Raw Score	Scale Score	Performance Level
86	100	5	57	81	4	28	66	3
85	99	5	56	81	4	27	65	3
84	97	5	55	80	4	26	64	2
83	96	5	54	80	4	25	62	2
82	95	5	53	80	4	24	61	2
81	94	5	52	80	4	23	60	2
80	93	5	51	79	3	22	58	2
79	92	5	50	79	3	21	57	2
78	91	5	49	79	3	20	55	2
77	90	5	48	78	3	19	53	1
76	90	5	47	78	3	18	52	1
75	89	5	46	78	3	17	50	1
74	88	5	45	77	3	16	48	1
73	88	5	44	77	3	15	46	1
72	87	5	43	77	3	14	44	1
71	86	5	42	76	3	13	42	1
70	86	5	41	76	3	12	39	1
69	86	5	40	75	3	11	37	1
68	85	5	39	75	3	10	34	1
67	84	4	38	74	3	9	32	1
66	84	4	37	73	3	8	29	1
65	83	4	36	73	3	7	26	1
64	83	4	35	72	3	6	23	1
63	83	4	34	71	3	5	19	1
62	82	4	33	70	3	4	16	1
61	82	4	32	70	3	3	12	1
60	82	4	31	69	3	2	8	1
59	82	4	30	68	3	1	4	1
58	81	4	29	67	3	0	0	1

Examination
August 2017
Algebra I

HIGH SCHOOL MATH REFERENCE SHEET

Conversions

1 inch = 2.54 centimeters	1 cup = 8 fluid ounces
1 meter = 39.37 inches	1 pint = 2 cups
1 mile = 5280 feet	1 quart = 2 pints
1 mile = 1760 yards	1 gallon = 4 quarts
1 mile = 1.609 kilometers	1 gallon = 3.785 liters
	1 liter = 0.264 gallon
1 kilometer = 0.62 mile	1 liter = 1000 cubic centimeters
1 pound = 16 ounces	
1 pound = 0.454 kilogram	
1 kilogram = 2.2 pounds	
1 ton = 2000 pounds	

Formulas

Triangle	$A = \frac{1}{2}bh$
Parallelogram	$A = bh$
Circle	$A = \pi r^2$
Circle	$C = \pi d$ or $C = 2\pi r$

Formulas (continued)

General Prisms	$V = Bh$
Cylinder	$V = \pi r^2 h$
Sphere	$V = \frac{4}{3}\pi r^3$
Cone	$V = \frac{1}{3}\pi r^2 h$
Pyramid	$V = \frac{1}{3}Bh$
Pythagorean Theorem	$a^2 + b^2 = c^2$
Quadratic Formula	$x = \dfrac{-b \pm \sqrt{b^2 - 4ac}}{2a}$
Arithmetic Sequence	$a_n = a_1 + (n-1)d$
Geometric Sequence	$a_n = a_1 r^{n-1}$
Geometric Series	$S_n = \dfrac{a_1 - a_1 r^n}{1 - r}$ where $r \neq 1$
Radians	$1 \text{ radian} = \dfrac{180}{\pi} \text{ degrees}$
Degrees	$1 \text{ degree} = \dfrac{\pi}{180} \text{ radians}$
Exponential Growth/Decay	$A = A_0 e^{k(t - t_0)} + B_0$

PART I

Answer all 24 questions in this part. Each correct answer will receive 2 credits. No partial credit will be allowed. For each statement or question, write in the space provided the numeral preceding the word or expression that best completes the statement or answers the question. [48 credits]

1 A part of Jennifer's work to solve the equation $2(6x^2 - 3) = 11x^2 - x$ is shown below.

   Given: $2(6x^2 - 3) = 11x^2 - x$
   Step 1: $12x^2 - 6 = 11x^2 - x$

Which property justifies her first step?

(1) identity property of multiplication

(2) multiplication property of equality

(3) commutative property of multiplication

(4) distributive property of multiplication over subtraction   1 _____

2 Which value of x results in equal outputs for $j(x) = 3x - 2$ and $b(x) = |x + 2|$?

(1) -2 (3) $\dfrac{2}{3}$

(2) 2 (4) 4 2 _____

3 The expression $49x^2 - 36$ is equivalent to

(1) $(7x - 6)^2$

(2) $(24.5x - 18)^2$

(3) $(7x - 6)(7x + 6)$

(4) $(24.5x - 18)(24.5x + 18)$ 3 _____

4 If $f(x) = \dfrac{1}{2}x^2 - \left(\dfrac{1}{4}x + 3\right)$, what is the value of $f(8)$?

(1) 11 (3) 27

(2) 17 (4) 33 4 _____

5 The graph below models the height of a remote-control helicopter over 20 seconds during flight.

Over which interval does the helicopter have the *slowest* average rate of change?

(1) 0 to 5 seconds   (3) 10 to 15 seconds

(2) 5 to 10 seconds   (4) 15 to 20 seconds   5 _____

6 In the functions $f(x) = kx^2$ and $g(x) = |kx|$, k is a positive integer.

If k is replaced by $\frac{1}{2}$, which statement about these new functions is true?

(1) The graphs of both $f(x)$ and $g(x)$ become wider.

(2) The graph of $f(x)$ becomes narrower and the graph of $g(x)$ shifts left.

(3) The graphs of both $f(x)$ and $g(x)$ shift vertically.

(4) The graph of $f(x)$ shifts left and the graph of $g(x)$ becomes wider.   6 _____

7 Wenona sketched the polynomial $P(x)$ as shown on the axes below.

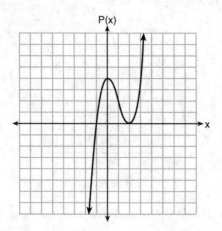

Which equation could represent $P(x)$?

(1) $P(x) = (x + 1)(x - 2)^2$

(2) $P(x) = (x - 1)(x + 2)^2$

(3) $P(x) = (x + 1)(x - 2)$

(4) $P(x) = (x - 1)(x + 2)$ 7 _____

8 Which situation does *not* describe a causal relationship?

(1) The higher the volume on a radio, the louder the sound will be.

(2) The faster a student types a research paper, the more pages the research paper will have.

(3) The shorter the time a car remains running, the less gasoline it will use.

(4) The slower the pace of a runner, the longer it will take the runner to finish the race.   8 _____

9 A plumber has a set fee for a house call and charges by the hour for repairs. The total cost of her services can be modeled by $c(t) = 125t + 95$.

Which statements about this function are true?

I.   A house call fee costs $95.

II.   The plumber charges $125 per hour.

III. The number of hours the job takes is represented by t.

(1) I and II, only   (3) II and III, only

(2) I and III, only   (4) I, II, and III   9 _____

10 What is the domain of the relation shown below?

$$\{(4, 2), (1, 1), (0, 0), (1, -1), (4, -2)\}$$

(1) $\{0, 1, 4\}$

(2) $\{-2, -1, 0, 1, 2\}$

(3) $\{-2, -1, 0, 1, 2, 4\}$

(4) $\{-2, -1, 0, 0, 1, 1, 1, 2, 4, 4\}$ 10 _____

11 What is the solution to the inequality $2 + \frac{4}{9}x \geq 4 + x$?

(1) $x \leq -\frac{18}{5}$ (3) $x \leq \frac{54}{5}$

(2) $x \geq -\frac{18}{5}$ (4) $x \geq \frac{54}{5}$ 11 _____

12 Konnor wants to burn 250 Calories while exercising for 45 minutes at the gym. On the treadmill, he can burn 6 Cal/min. On the stationary bike, he can burn 5 Cal/min.

If t represents the number of minutes on the treadmill and b represents the number of minutes on the stationary bike, which expression represents the number of Calories that Konnor can burn on the stationary bike?

(1) b (3) $45 - b$

(2) $5b$ (4) $250 - 5b$ 12 _____

13 Which value of x satisfies the equation $\frac{5}{6}\left(\frac{3}{8} - x\right) = 16$?

   (1) -19.575 (3) -16.3125

   (2) -18.825 (4) -15.6875 13 _____

14 If a population of 100 cells triples every hour, which function represents $p(t)$, the population after t hours?

   (1) $p(t) = 3(100)t$ (3) $p(t) = 3t + 100$

   (2) $p(t) = 100(3)t$ (4) $p(t) = 100t + 3$ 14 _____

15 A sequence of blocks is shown in the diagram below.

This sequence can be defined by the recursive function $a_1 = 1$ and $a_n = a_{n-1} + n$. Assuming the pattern continues, how many blocks will there be when $n = 7$?

   (1) 13   (3) 28

   (2) 21   (4) 36   15 _____

16 Mario's \$15,000 car depreciates in value at a rate of 19% per year. The value, V, after t years can be modeled by the function $V = 15,000(0.81)^t$. Which function is equivalent to the original function?

   (1) $V = 15,000(0.9)^{9t}$ (3) $V = 15,000(0.9)^{\frac{t}{9}}$

   (2) $V = 15,000(0.9)^{2t}$ (4) $V = 15,000(0.9)^{\frac{t}{2}}$ 16 _____

17 The highest possible grade for a book report is 100. The teacher deducts 10 points for each day the report is late.

Which kind of function describes this situation?

   (1) linear   (3) exponential growth

   (2) quadratic   (4) exponential decay   17 _____

18 The function $h(x)$, which is graphed below, and the function
   $g(x) = 2|x + 4| - 3$ are given.

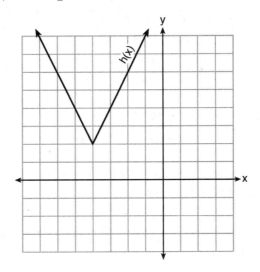

Which statements about these functions are true?

I. $g(x)$ has a lower minimum value than $h(x)$.

II. For all values of x, $h(x) < g(x)$.

III. For any value of x, $g(x) \neq h(x)$.

(1) I and II, only   (3) II and III, only

(2) I and III, only   (4) I, II, and III   18 ____

19 The zeros of the function $f(x) = 2x^3 + 12x - 10x^2$ are

(1) $\{2, 3\}$ (3) $\{0, 2, 3\}$

(2) $\{-1, 6\}$ (4) $\{0, -1, 6\}$ 19 ____

20 How many of the equations listed below represent the line passing
    through the points $(2, 3)$ and $(4, -7)$?

$$5x + y = 13$$
$$y + 7 = -5(x - 4)$$
$$y = -5x + 13$$
$$y - 7 = 5(x - 4)$$

(1) 1   (3) 3
(2) 2   (4) 4   20 _____

21 The Ebola virus has an infection rate of 11% per day as compared
    to the SARS virus, which has a rate of 4% per day.

    If there were one case of Ebola and 30 cases of SARS initially
    reported to authorities and cases are reported each day, which
    statement is true?

    (1) At day 10 and day 53 there are more Ebola cases.
    (2) At day 10 and day 53 there are more SARS cases.
    (3) At day 10 there are more SARS cases, but at day 53 there
         are more Ebola cases.
    (4) At day 10 there are more Ebola cases, but at day 53 there
         are more SARS cases.   21 _____

22 The results of a linear regression are shown below.

$$y = ax + b$$
$$a = -1.15785$$
$$b = 139.3171772$$
$$r = -0.896557832$$
$$r^2 = 0.8038159461$$

    Which phrase best describes the relationship between x and y?

    (1) strong negative correlation
    (2) strong positive correlation
    (3) weak negative correlation
    (4) weak positive correlation   22 _____

23 Abigail's and Gina's ages are consecutive integers. Abigail is younger than Gina and Gina's age is represented by x. If the difference of the square of Gina's age and eight times Abigail's age is 17, which equation could be used to find Gina's age?

(1) $(x + 1)^2 - 8x = 17$ (3) $x^2 - 8(x + 1) = 17$

(2) $(x - 1)^2 - 8x = 17$ (4) $x^2 - 8(x - 1) = 17$ 23 ____

24 Which system of equations does *not* have the same solution as the system below?

$$4x + 3y = 10$$
$$-6x - 5y = 16$$

(1) $-12x - 9y = -30$ (3) $24x + 18y = 60$
     $12x + 10y = 32$ $-24x - 20y = -64$

(2) $20x + 15y = 50$ (4) $40x + 30y = 100$
     $-18x - 15y = -48$ $36x - 30y = -96$ 24 ____

PART II

Answer all 8 questions in this part. Each correct answer will receive 2 credits. Clearly indicate the necessary steps, including appropriate formula substitutions, diagrams, graphs, charts, etc. For all questions in this part, a correct numerical answer with no work shown will receive only 1 credit.   [16 credits]

25 A teacher wrote the following set of numbers on the board:

$$a = \sqrt{20} \qquad b = 2.5 \qquad c = \sqrt{225}$$

Explain why $a + b$ is irrational, but $b + c$ is rational.

26 Determine and state whether the sequence 1, 3, 9, 27, . . . displays exponential behavior. Explain how you arrived at your decision.

27 Using the formula for the volume of a cone, express r in terms of V, h, and π.

28 The graph below models the cost of renting video games with a membership in Plan A and Plan *B*.

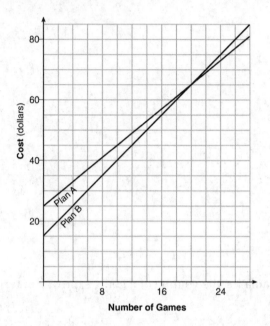

Explain why Plan *B* is the better choice for Dylan if he only has $50 to spend on video games, including a membership fee.

Bobby wants to spend $65 on video games, including a membership fee. Which plan should he choose? Explain your answer.

29 Samantha purchases a package of sugar cookies. The nutrition label states that each serving size of 3 cookies contains 160 Calories. Samantha creates the graph below showing the number of cookies eaten and the number of Calories consumed.

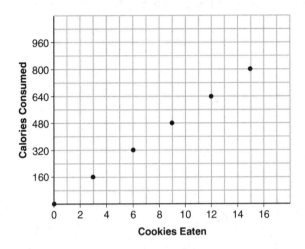

Explain why it is appropriate for Samantha to draw a line through the points on the graph.

30 A two-inch-long grasshopper can jump a horizontal distance of 40 inches. An athlete, who is five feet nine, wants to cover a distance of one mile by jumping. If this person could jump at the same ratio of body-length to jump-length as the grasshopper, determine, to the *nearest jump*, how many jumps it would take this athlete to jump one mile.

31 Write the expression $5x + 4x^2(2x + 7) - 6x^2 - 9x$ as a polynomial in standard form.

32 Solve the equation $x^2 - 6x = 15$ by completing the square.

PART III

Answer all 4 questions in this part. Each correct answer will receive 4 credits. Clearly indicate the necessary steps, including appropriate formula substitutions, diagrams, graphs, charts, etc. For all questions in this part, a correct numerical answer with no work shown will receive only 1 credit. [16 credits]

33 Loretta and her family are going on vacation. Their destination is 610 miles from their home. Loretta is going to share some of the driving with her dad. Her average speed while driving is 55 mph and her dad's average speed while driving is 65 mph.

The plan is for Loretta to drive for the first 4 hours of the trip and her dad to drive for the remainder of the trip. Determine the number of hours it will take her family to reach their destination.

After Loretta has been driving for 2 hours, she gets tired and asks her dad to take over. Determine, to the *nearest tenth of an hour*, how much time the family will save by having Loretta's dad drive for the remainder of the trip.

34 The heights, in feet, of former New York Knicks basketball players are listed below.

6.4	6.9	6.3	6.2	6.3	6.0	6.1	6.3	6.8	6.2
6.5	7.1	6.4	6.3	6.5	6.5	6.4	7.0	6.4	6.3
6.2	6.3	7.0	6.4	6.5	6.5	6.5	6.0	6.2	

Using the heights given, complete the frequency table below.

Interval	Frequency
6.0–6.1	
6.2–6.3	
6.4–6.5	
6.6–6.7	
6.8–6.9	
7.0–7.1	

Question 34 is continued on the next page.

Question 34 continued

Based on the frequency table created, draw and label a frequency histogram on the grid below.

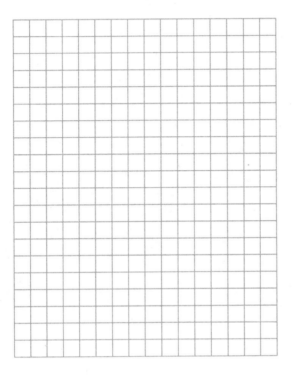

Determine and state which interval contains the upper quartile. Justify your response.

35 Solve the following system of inequalities graphically on the grid below and label the solution S.

$$3x + 4y > 20$$
$$x < 3y - 18$$

Is the point $(3, 7)$ in the solution set? Explain your answer.

36 An Air Force pilot is flying at a cruising altitude of 9000 feet and is forced to eject from her aircraft. The function $h(t) = -16t^2 + 128t + 9000$ models the height, in feet, of the pilot above the ground, where t is the time, in seconds, after she is ejected from the aircraft.

Determine and state the vertex of $h(t)$. Explain what the second coordinate of the vertex represents in the context of the problem.

After the pilot was ejected, what is the maximum number of feet she was above the aircraft's cruising altitude? Justify your answer.

PART IV

Answer the question in this part. A correct answer will receive 6 credits. Clearly indicate the necessary steps, including appropriate formula substitutions, diagrams, graphs, charts, etc. A correct numerical answer with no work shown will receive only 1 credit. [6 credits]

37 Zeke and six of his friends are going to a baseball game. Their combined money totals $28.50. At the game, hot dogs cost $1.25 each, hamburgers cost $2.50 each, and sodas cost $0.50 each. Each person buys one soda. They spend all $28.50 on food and soda.

Write an equation that can determine the number of hot dogs, x, and hamburgers, y, Zeke and his friends can buy.

Question 37 is continued on the next page.

Question 37 continued

Graph your equation on the grid below.

Determine how many different combinations, including those combinations containing zero, of hot dogs and hamburgers Zeke and his friends can buy, spending all $28.50. Explain your answer.

Answers
August 2017
Algebra I

Answer Key

PART I

1. (4)	5. (2)	9. (4)	13. (2)	17. (1)	21. (3)
2. (2)	6. (1)	10. (1)	14. (2)	18. (2)	22. (1)
3. (3)	7. (1)	11. (1)	15. (3)	19. (3)	23. (4)
4. (3)	8. (2)	12. (2)	16. (2)	20. (3)	24. (4)

PART II

25. Rational + Irrational = Irrational; Rational + Rational = Rational

26. Yes, the common ratio is 3.

27. $r = \sqrt{\dfrac{3V}{\pi h}}$

28. For \$50, Dylan can get 14 games with B compared to 12 games with A. For \$65, either plan gets Bobby 20 games.

29. She does not have to eat cookies 3 at a time. She can also eat a fraction of a cookie.

30. 46 jumps

31. $8x^3 + 22x^2 - 4x$

32. $x = 3 \pm \sqrt{24}$ or $x = 3 \pm 2\sqrt{6}$

PART III

33. 10 hours, 0.3 hours

34. See diagram, the top quartile is in the 6.4–6.5 interval

35. See diagram, (3, 7) is not in the solution set

36. (4, 9,256), 256 feet

PART IV

37. $1.25x + 2.50y = 25$, 11 different combinations

In **Parts II–IV**, you are required to show how you arrived at your answers. For sample methods of solutions, see the *Answers Explained* section.

Answers and Explanations

PART I

1. The distributive property of multiplication over subtraction says that an expression in the form $a \cdot (b - c)$ is equivalent to the expression $a \cdot b - a \cdot c$. In the solution, this property is applied to change $2(6x^3 - 3)$ to $2 \cdot 6x^3 - 2 \cdot 3 = 12x^3 - 6$.

   The correct choice is (**4**).

2. Test each of the answer choices to see which one makes $j(x) = b(x)$.

   • Choice (1):

$$j(-2) = 3 \cdot (-2) - 2$$
$$= -6 - 2$$
$$= -8$$

$$b(-2) = \left| -2 + 2 \right|$$
$$= \left| 0 \right|$$
$$= 0$$

   • Choice (2):

$$j(2) = 3 \cdot 2 - 2$$
$$= 6 - 2$$
$$= 4$$

$$b(2) = \left| 2 + 2 \right|$$
$$= \left| 4 \right|$$
$$= 4$$

   The correct choice is (**2**).

3. The difference of perfect squares factoring pattern says that an expression of the form $a^2 - b^2$ is equivalent to the expression $(a - b)(a + b)$. The expression $49x^2 - 36$ can be rewritten as $(7x)^2 - 6^2$. Since this is a difference of perfect squares, it is equal to $(7x - 6)(7x + 6)$.

The correct choice is (**3**).

4. Replace each x with the number 8 and simplify:

$$f(8) = \frac{1}{2} \cdot 8^2 - \left(\frac{1}{4} \cdot 8 + 3\right)$$
$$= \frac{1}{2} \cdot 64 - (2 + 3)$$
$$= 32 - 5$$
$$= 27$$

The correct choice is (**3**).

5. The average rate of change over an interval is the slope of the line segment that joins the endpoints of that interval. The slowest rate of change corresponds to the interval whose slope has the smallest absolute value. Even without doing any calculations, it can be seen that the interval from 5 to 10 seconds has the slope with the smallest absolute value since a line segment joining $(5, 19)$ and $(10, 20)$ is closest to being a horizontal line segment.

The average rate of change can also be calculated by using the slope formula $m = \dfrac{y_2 - y_1}{x_2 - x_1}$ for each of the intervals.

- Choice (1):

$$m = \frac{19 - 46}{5 - 0}$$
$$= \frac{-27}{5}$$
$$= -5\frac{2}{5}$$

The absolute value is $5\frac{2}{5}$.

- Choice (2):

$$m = \frac{20 - 19}{10 - 5}$$

$$= \frac{1}{5}$$

The absolute value is $\frac{1}{5}$.

- Choice (3):

$$m = \frac{14 - 20}{15 - 10}$$

$$= \frac{-6}{5}$$

$$= -1\frac{1}{5}$$

The absolute value is $1\frac{1}{5}$.

- Choice (4):

$$m = \frac{17 - 14}{20 - 15}$$

$$= \frac{3}{5}$$

The absolute value is $\frac{3}{5}$.

The correct choice is (**2**).

6. If you choose an initial integer value for k, such as $k = 2$, you can use a graphing calculator to compare the graphs of $f(x) = 2x^2$ to $f(x) = \frac{1}{2}x^2$ and the graphs of $g(x) = |\,2x\,|$ to $g(x) = \left|\,\frac{1}{2}x\,\right|$.

For the TI-84:

For the TI-Nspire:

When k is replaced with $\frac{1}{2}$, each of the graphs becomes wider.

The correct choice is (**1**).

7. Each of the answer choices can be graphed on the graphing calculator. When choice (1) is graphed, it looks like the graph from the question.

For the TI-84:

For the TI-Nspire:

Even without the graphing calculator, it can be seen that the graph has an x-intercept at -1. So $(x + 1)$ would be one factor. The graph also has a double root at $x = 2$ since the graph bounces off the axis rather than passing through the axis. So there will be two $(x - 2)$ factors.

The correct choice is (**1**).

8. A causal relationship is one where one event causes the other to happen.

   - Choice (1): Raising the volume does cause the radio to become louder.
   - Choice (2): How fast the student types does not matter. If the paper is five pages long, the student will complete typing the paper more quickly if he or she types faster, but the paper will not be longer.
   - Choice (3): Assuming that the car uses the same amount of gasoline each minute it runs, the shorter it remains running, the less gasoline the car will use.
   - Choice (4): Running slower does cause a runner to take more time to complete the race.

   There is an argument to be made that choice (2) could be considered causal if, for example, the student decides to spend exactly two hours on the paper and to type for that amount of time. In that case, the faster typist will have completed more pages after two hours. Compared to the other choices, though, choice (2) is still the best choice.

   The correct choice is (**2**).

9. One way to determine what the components of this equation represent is to calculate the values of $c(0)$, $c(1)$, and $c(2)$:

$$c(0) = 125 \cdot 0 + 95$$
$$= 0 + 95$$
$$= 95$$
$$c(1) = 125 \cdot 1 + 95$$
$$= 125 + 95$$
$$= 220$$
$$c(2) = 125 \cdot 2 + 95$$
$$= 250 + 95$$
$$= 345$$

   So when $t = 0$, the function evaluates to 95. This means that \$95 must be the cost to get the plumber to come to the house even without spending any time doing the repairs. When $t = 1$, the fee increases by \$125. So 125 must be the price per hour.

   In general, when a real-world scenario has the form $y = mx + b$, the b represents the fixed costs and the m represents the cost increase for every time x increases by 1.

   The correct choice is (**4**).

10. The domain of a set of ordered pairs is a set containing each unique x-coordinate from each of the ordered pairs. The x-coordinates of the five ordered pairs are 4, 1, 0, 1, and 4. Since a set does not contain repeated numbers, this domain can be described as $\{0, 1, 4\}$.

The correct choice is (**1**).

11. Start by solving the inequality the way you would solve an equality:

$$2 + \frac{4}{9}x \geq 4 + x$$

$$-x = -x$$

$$2 + \frac{4}{9}x - \frac{9}{9}x \geq 4$$

$$2 - \frac{5}{9}x \geq 4$$

$$-2 = -2$$

$$-\frac{5}{9}x \geq 2$$

For the last step, multiply both sides of the equation by $-\frac{9}{5}$. With inequalities, you have to switch the direction of the inequality sign anytime you multiply or divide both sides of the inequality by a negative:

$$\left(-\frac{9}{5}\right)\left(-\frac{5}{9}\right)x \leq \left(-\frac{9}{5}\right)2$$

$$x \leq -\frac{18}{5}$$

The correct choice is (**1**).

12. If you were trying to solve for the number of minutes on each of the exercise machines, you would need to solve a system of equations:

$$t + b = 45$$

$$6t + 5b = 250$$

However, this question does not ask you to solve for t and b. It just asks how many calories Konnor burns on the bike. If he spends b hours on the bike and burns 5 Cal/min each minute on the bike, he burns $5b$ calories on the bike. The other information in the question is irrelevant.

The correct choice is (**2**).

13. Solve the equation for x:

$$\frac{5}{6}\left(\frac{3}{8} - x\right) = 16$$

$$\left(\frac{5}{6}\right)\left(\frac{3}{8}\right) - \frac{5}{6}x = 16$$

$$\frac{15}{48} - \frac{5}{6}x = 16$$

$$-\frac{15}{48} = -\frac{15}{48}$$

$$-\frac{5}{6}x = 16 - \frac{15}{48}$$

$$= 15.6875$$

$$\left(-\frac{6}{5}\right)\left(-\frac{6}{5}x\right) = \left(-\frac{6}{5}\right) \cdot 15.6875$$

$$x = -18.825$$

The correct choice is (**2**).

14. The population of cells forms the sequence 100, 300, 900, 2,700, ...

If you substitute 0, 1, and 2 into the functions given in the choices, choice (2) becomes

$$p(0) = 100(3)^0$$
$$= 100 \cdot 1$$
$$= 100$$

$$p(1) = 100(3)^1$$
$$= 100 \cdot 3$$
$$= 300$$

$$p(0) = 100(3)^2$$
$$= 100 \cdot 9$$
$$= 900$$

An alternate way of answering this is to notice that since the numbers increase exponentially, the function must be of the form $p(t) = a \cdot b^t$. In a function of this form, a represents the initial value and b represents the growth factor. The initial value is 100, and the growth factor is 3. So the function is $p(t) = 100(3)^t$.

The correct choice is (**2**).

15. The given values of the sequence are $a_1 = 1$, $a_2 = 3$, $a_3 = 6$, and $a_4 = 10$. To get the next three values, substitute $n = 5$, 6, and 7 into the recursive function definition:

$$a_5 = a_{5-1} + 5$$
$$= a_4 + 5$$
$$= 10 + 5$$
$$= 15$$
$$a_6 = a_{6-1} + 6$$
$$= a_5 + 6$$
$$= 15 + 6$$
$$= 21$$
$$a_7 = a_{7-1} + 7$$
$$= a_6 + 7$$
$$= 21 + 7$$
$$= 28$$

The correct choice is (**3**).

16. The simplest way to solve this question is to substitute $t = 1$ into each of the answer choices to see which gives the same answer as $V = 15,000(0.81)^1 = 15,000 \cdot 0.81 = 12,150$.

When you substitute $t = 1$ into choice (2), it becomes:

$$V = 15,000(0.9)^{2.1}$$
$$= 15,000(0.9)^2$$
$$= 15,000 \cdot 0.81$$
$$= 12,150$$

Another way to see that choice (2) is equivalent to the original equation is to use the property of exponents that $x^{ab} = (x^a)^b$:

$$15,000 \cdot (0.9)^{2t} = 15,000\left(0.9^2\right)^t$$
$$= 15,000(0.81)^t$$

The correct choice is (**2**).

17. If you make a graph where for each point the x-coordinate is the number of days late and the y-coordinate is the highest possible grade, it will look like this:

Since the points all fall on a line, this is a linear relationship.

The correct choice is (**1**).

18. When you graph $g(x)$ on the same set of axes as $h(x)$, they look like this:

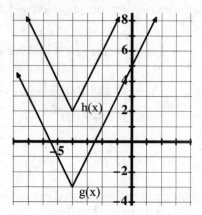

Using the graphing calculator, plot function g. It becomes the same as the graph of h but shifted down 5 units.

For the TI-84:

For the TI-Nspire:

- Statement I: The minimum value of $g(x)$ is -3, while the minimum value of $h(x)$ is 2. So the minimum value of $g(x)$ is lower than the minimum value of $h(x)$.

- Statement II: For every x-value, the y-coordinate of the graph of $h(x)$ is above the y-coordinate of the graph of $g(x)$. So it is not true that for all (or any) values of x that $h(x) < g(x)$.

- Statement III: Since $g(x)$ is the same as $h(x)$ but just shifted down 5 units, there is no x-value where $g(x) = h(x)$. So it is true that for any value of x, $g(x) \neq h(x)$.

Statements I and III are both true.

The correct choice is (**2**).

19. The zeros of a function are the values that make the function evaluate to zero. To calculate the zeros of f, find all the solutions to the equation $0 = 2x^3 + 12x - 10x^2$:

$$0 = 2x^3 + 12x - 10x^2$$
$$0 = 2x\left(x^2 + 6 - 5x\right)$$
$$= 2x\left(x^2 - 5x + 6\right)$$
$$= 2x(x - 2)(x - 3)$$

For the product to be equal to zero, $2x$, $x - 2$, or $x - 3$ must equal zero:

$2x = 0$	$x - 2 = 0$	$x - 3 = 0$
$\dfrac{2x}{2} = \dfrac{0}{2}$ or	$+2 = +2$ or	$+3 = +3$
$x = 0$	$x = 2$	$x = 3$

The solution set is $\{0, 2, 3\}$.

If you divided both sides of the equation by x, you would have "lost" the $x = 0$ solution and found only $\{2, 3\}$ as your solution, which is incorrect.

The correct choice is (**3**).

20. To find the equation of a line through two given points, first find the slope of the line. In this case, the slope is

$$m = \frac{-7 - 3}{4 - 2}$$
$$= \frac{-10}{2}$$
$$= -5$$

To get the equation of the line in slope-intercept form, pick one of the two points. Insert its coordinates and the slope you just calculated into the equation $y = mx + b$. Then solve for b:

$$y = mx + b$$
$$3 = -5.2 + b$$
$$3 = -10 + b$$
$$+10 = +10$$
$$13 = b$$

In slope-intercept form, the equation of the line is $y = -5x + 13$.

Of the four equations listed, this exactly matches the third. However, you must rearrange the others and then compare them to $y = -5x + 13$:

- First equation:

$$5x + y = 13$$
$$-5x = -5x$$
$$y = -5x + 13$$
$$\checkmark$$

- Second equation:

$$y + 7 = -5(x - 4)$$
$$y + 7 = -5x + 20$$
$$-7 = -7$$
$$y = -5x + 13$$
$$\checkmark$$

- Fourth equation:

$$y - 7 = 5(x - 4)$$
$$y - 7 = 5x - 20$$
$$+7 = +7$$
$$y = 5x - 13$$

NO

Of the four equations, the first three are equivalent to $y = -5x + 13$ while the fourth is not. Three of the four equations, then, represent the line passing through the two points.

The correct choice is (**3**).

21. The formula for exponential growth is $y = a(1 + r)^x$, where a is the starting value and r is the growth rate.

For the Ebola virus the formula is

$$y = 1(1 + 0.11)^x$$
$$= 1(1.11)^x$$

For the SARS virus the formula is:

$$y = 30(1 + .04)^x$$
$$= 30(1.04)^x$$

At day 10, the number of Ebola cases is

$$y = 1(1.11)^{10}$$
$$= 2.84$$
$$\approx 3$$

At day 10, the number of SARS cases is

$$y = 30(1.04)^{10}$$
$$= 44.4$$
$$\approx 44$$

At day 53, the number of Ebola cases is

$$y = 1(1.11)^{53}$$
$$= 252.4$$
$$\approx 252$$

At day 53, the number of SARS cases is

$$y = 30(1.04)^{53}$$
$$= 239.8$$
$$\approx 240$$

At day 10, there are more SARS cases. At day 53, there are more Ebola cases.

The correct choice is (**3**).

22. This information tells you that the equation for the line of best fit for a scatter plot is $y = -1.15785x + 139.3171772$. Since the slope of the line is negative, the relationship between x and y is negative. So eliminate choices (3) and (4).

The strength of the correlation is based on the r^2-value. The closer that is to 1, the stronger the correlation. Since 0.8038159461 is pretty close to 1, this can be considered a strong negative correlation.

The correct choice is (**1**).

23. Gina's age is represented by x. If Abigail is younger than Gina and their ages are consecutive integers, Gina must be one year younger than Abigail. So Gina's age can be represented by $x - 1$.

    The square of Gina's age is x^2.
    Eight times Abigail's age is $8(x - 1)$.

    If the difference between these is 17, an equation that could be used to find Gina's age is $x^2 - 8(x - 1) = 17$.

    The correct choice is (**4**).

24. In a system of equations, if you multiply both sides of one (or both) of the equations by a constant, the new system will have the same solution as the original system. Look at each of the four choices.

    - Choice (1): This is what you would get if you multiplied both sides of the top equation by -3 and both sides of the bottom equation by -2.
    - Choice (2): This is what you would get if you multiplied both sides of the top equation by 5 and both sides of the bottom equation by 3.
    - Choice (3): This is what you would get if you multiplied both sides of the top equation by 6 and both sides of the bottom equation by 4.
    - Choice (4): Although you could get this top equation by multiplying both sides of the original top equation by 10, there is no way to get the bottom equation by multiplying both sides of the original bottom equation by the same number. If you multiply the left side of the bottom equation by -6 and the right side by $+6$, the new system of equations will have a different solution than the original system.

    The correct choice is (**4**).

PART II

25. a is irrational because 20 is not a perfect square. b is rational because it is a terminating decimal. Even though c is a square root, it is rational because 225 is a perfect square and $\sqrt{225} = 15$.

$a + b$ is irrational because an irrational plus a rational is always irrational. $b + c$ is rational because a rational plus a rational is always rational.

Note: Sometimes an irrational plus an irrational is irrational while other times an irrational plus an irrational is a rational number.

26. A sequence in which each term is equal to the previous term multiplied by some constant exhibits exponential behavior. In this case, the geometric sequence 1, 3, 9, 27, . . . has an initial term of 1 and a constant ratio of 3.

27. The formula for the volume of a cone is given in the reference sheet: $V = \frac{1}{3}\pi r^2 h$. To express r in terms of the other variables, treat the V and h as numbers and "solve" for r:

$$V = \frac{1}{3}\pi r^2 h$$

$$3 \cdot V = 3 \cdot \frac{1}{3}\pi r^2 h$$

$$3V = \pi r^2 h$$

$$\frac{3V}{\pi h} = \frac{\pi r^2 h}{\pi h}$$

$$\frac{3V}{\pi h} = r^2$$

$$\sqrt{\frac{3V}{\pi h}} = \sqrt{r^2}$$

$$\sqrt{\frac{3V}{\pi h}} = r$$

You should not include a $\pm$ symbol in front of the radical because the radius of a cone must be a positive number.

28. If you draw a horizontal line at the $50 mark on the vertical axis, you will see that the line for Plan *A* contains a point at approximately (12.25, 50) while the line for Plan *B* contains a point at (14, 50). This means that with Plan *A*, you can get only 12 games for $50 while with Plan *B* you can get 14. Plan *B* is a better choice for Dylan since he can get two additional games.

Because both lines contain the point (20, 65), it doesn't matter which plan Bobby chooses. With either plan, he can get 20 games for $65.

29. It is appropriate for Samantha to draw a line through the points on the graph because it is possible to eat a number of cookies that is not a multiple of 3. You can have 1 cookie or 2 cookies or even a half a cookie. The number of calories is proportional to the number of cookies. So one cookie has one-third the number of calories as 3 cookies, and half of a cookie has half as many calories as one cookie.

In mathematical terms, we say that the domain of the graph is all non-negative real numbers.

30. First set up a proportion to find the number of inches that the athlete can jump. Convert the athlete's height to inches so everything is expressed in the same unit:

$$5 \text{ feet } 9 \text{ inches} = (12 \times 5) + 9$$
$$= 60 + 9$$
$$= 69 \text{ inches}$$

$$\frac{2 \text{ in.}}{40 \text{ in.}} = \frac{69 \text{ in.}}{x \text{ in.}}$$
$$2x = 40 \cdot 69 = 2760$$
$$\frac{2x}{2} = \frac{2760}{2}$$
$$x = 1380$$

The athlete can jump 1,380 inches each jump.

To calculate the number of jumps the athlete needs to accomplish to make a mile, divide the number of inches in a mile by 1,380:

$$\text{Feet in a mile} = 5,280$$
$$\text{Inches in a mile} = 12 \times 5,280$$
$$= 63,360$$

$$63,360 \div 1,380 = 45.9$$
$$\approx 46 \text{ jumps}$$

31. First use the distributive property, and then combine the like terms:

$$5x + 4x^2(2x + 7) - 6x^2 - 9x$$
$$5x + 8x^3 + 28x^2 - 6x^2 - 9x$$
$$8x^3 + 28x^2 - 6x^2 + 5x - 9x$$
$$8x^3 + 22x^2 - 4x$$

32. To complete the square with an equation like $x^2 + bx = c$, add $\left(\dfrac{b}{2}\right)^2$ to both sides of the equation:

$$x^2 - 6x = 15$$

$$+\left(\dfrac{-6}{2}\right)^2 = +\left(\dfrac{-6}{2}\right)^2$$

$$x^2 - 6x + 9 = 15 + 9$$

$$x^2 - 6x + 9 = 24$$

The left side of this equation is now a perfect square and can be factored:

$$x^2 - 6x + 9 = 24$$

$$(x - 3)^2 = 24$$

Take the square root of both sides of the equation, and isolate the x-term:

$$(x - 3)^2 = 24$$

$$\sqrt{(x - 3)^2} = \pm\sqrt{24}$$

$$x - 3 = \pm\sqrt{24}$$

$$+3 = +3$$

$$x = 3 \pm \sqrt{24}$$

This can be simplified to $x = 3 \pm 2\sqrt{6}$.

PART III

33. In 4 hours, Loretta drives $4 \times 55 = 220$ miles. This leaves $610 - 220 = 390$ miles. At 65 mph, 390 miles can be traveled in $390 \div 65 = 6$ hours. This, together with Loretta's 4 hours of driving, is a total of 10 hours.

    In 2 hours, Loretta drives $2 \times 55 = 110$ miles. This leaves $610 - 110 = 500$ miles. At 65 mph, 500 miles can be traveled in $500 \div 65 = 7.7$ hours. This, together with Loretta's 2 hours of driving, is a total of approximately 9.7 hours. This is 0.3 hours faster than the original plan.

34.

Interval	Frequency
6.0–6.1	3
6.2–6.3	10
6.4–6.5	11
6.6–6.7	0
6.8–6.9	2
7.0–7.1	3

A histogram is like a bar graph with no spaces between the bars. In a histogram you can choose the size of the intervals. Using an interval of 0.2, you get a histogram like this

With 29 numbers, the 15th is the median. To get the top quartile, find the median of the top 14 numbers. Since the 7th and 8th largest numbers are both 6.5, the top quartile is 6.5, which is contained in the 6.4–6.5 interval.

35. Start by changing the inequalites into slope-intercept form.

First inequality:

$$3x + 4y > 20$$
$$-3x = -3x$$
$$4y > -3x + 20$$
$$\frac{4y}{4} > \frac{-3x + 20}{4}$$
$$y > -\frac{3}{4}x + 5$$

Second inequality:

$$x < 3y - 18$$
$$+18 = +18$$
$$x + 18 < 3y$$
$$\frac{x + 18}{3} < \frac{3y}{3}$$
$$\frac{1}{3}x + 6 < y$$
$$y > \frac{1}{3}x + 6$$

Graph the boundary line for the first inequality like $y = -\frac{3}{4}x + 5$ but as a dotted line because of the $>$ sign. If there was a $\geq$ sign, the graph would need to be a solid line.

To decide which side of the line to shade, check if the ordered pair $(0, 0)$ makes the inequality true:

$$0 > -\frac{3}{4} \cdot 0 + 5$$
$$0 > 0 + 5$$
$$0 > 5$$
$$\textbf{NO}$$

Since it is not true that $0 > 5$, shade the side of the line that does not contain $(0, 0)$.

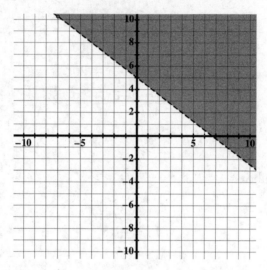

Graph the boundary line for the second inequality like $y = \frac{1}{3}x + 6$ but as a dotted line because of the $>$ sign.

To decide which side of the line to shade, check if the ordered pair $(0, 0)$ makes the inequality true:

$$0 > \frac{1}{3} \cdot 0 + 6$$
$$0 > 0 + 6$$
$$0 > 6$$
NO

Since it is not true that $0 > 6$, shade the side of the line that does not contain $(0, 0)$.

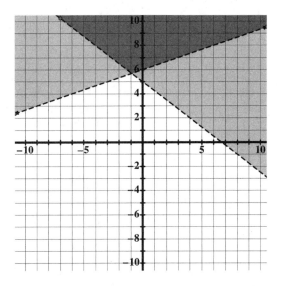

This can also be done with the graphing calculator.

For the TI-84:

For the TI-Nspire:

The point $(3, 7)$ is not in the solution set because it is on the dotted boundary line. You can also verify this by checking if $(3, 7)$ makes both inequalities true. Since $3 < 3 \cdot 7 - 18$ is not true (they are equal), the ordered pair $(3, 7)$ does not satisfy the second inequality. So $(3, 7)$ cannot be part of the solution set.

36. One way to find the vertex of $h(x)$ is to graph the function on the graphing calculator and to use the maximum command to find the coordinates of the vertex.

For the TI-84:

For the TI-Nspire:

The x-coordinate of the vertex of a quadratic function (a parabola) can also be found by using the formula $x = \dfrac{-b}{2a}$. For this function, $x = \dfrac{-b}{2a} = \dfrac{-128}{2(-16)} = 4$. The y-coordinate of the vertex will be

$$h(4) = -16 \cdot 4^2 + 128 \cdot 4 + 9{,}000$$
$$= 9{,}256$$

The vertex of $h(t)$ is $(4, 9{,}256)$.

Since cruising altitude is 9,000 feet and the maximum height achieved is 9,256 feet, the answer is $9{,}256 - 9{,}000 = 256$ feet above the cruising altitude.

PART IV

37. Since there are 7 people and they each buy one soda, they spend $7 \times 0.50 =$ $3.50 on sodas. This leaves $28.50 - $3.50 = $25.00 for the hot dogs and hamburgers.

Since each hot dog costs $1.25 and there are x hot dogs, the total cost of the hotdogs is $1.25x$. Since each hamburger costs $2.50 and there are y hamburgers, the total cost of the hamburgers is $2.50y$. For the total cost of all the hot dogs and hamburgers to be $25.00, the equation is

$$1.25x + 2.50y = 25$$

To graph this equation, convert to slope-intercept form:

$$1.25x + 2.50y = 25$$
$$-1.25x = -1.25x$$
$$2.50y = -1.25x + 25$$
$$\frac{2.50y}{2.50} = \frac{-1.25x + 25}{2.50}$$
$$y = -\frac{1}{2}x + 10$$

Graph this line on the grid.

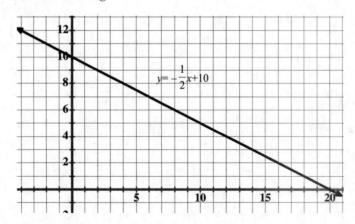

Since you cannot buy a fraction of a hot dog or a hamburger, find the coordinates of the points that have integers as both the x-coordinate and y-coordinate. There are 11 different combinations.

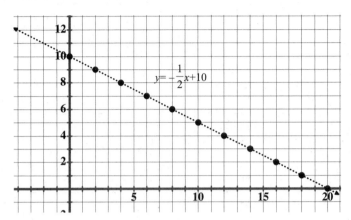

Hot Dogs	Hamburgers
20	0
18	1
16	2
14	3
12	4
10	5
8	6
6	7
4	8
2	9
0	10

Topic	Question Numbers	Number of Points	Your Points	Your Percentage
1. Polynomials	7, 31	$2 + 2 = 4$		
2. Properties of Algebra	1, 13, 27, 30	$2 + 2 + 2 + 2 = 8$		
3. Functions	2, 4, 5, 10, 17, 20, 19	$2 + 2 + 2 + 2 + 2 + 2 + 2 = 14$		
4. Creating and Interpreting Equations	9, 23, 33, 37	$2 + 2 + 4 + 6 = 14$		
5. Inequalities	11, 35	$2 + 4 = 6$		
6. Sequences and Series	15, 26	$2 + 2 = 4$		
7. Systems of Equations	24, 28	$2 + 2 = 4$		
8. Quadratic Equations and Factoring	3, 19, 32, 36	$2 + 2 + 2 + 4 = 10$		
9. Regression	22	2		
10. Exponential Equations	14, 16, 21	$2 + 2 + 2 = 6$		
11. Graphing	6, 18	$2 + 2 = 4$		
12. Statistics	8, 12, 24, 28, 34	$2 + 2 + 2 + 2 + 4 = 12$		
13. Number Properties	25	2		

HOW TO CONVERT YOUR RAW SCORE TO YOUR ALGEBRA I REGENTS EXAMINATION SCORE

The accompanying conversion chart must be used to determine your final score on the August 2017 Regents Examination in Algebra I. To find your final exam score, locate in the column labeled "Raw Score" the total number of points you scored out of a possible 86 points. Since partial credit is allowed in Parts II, III, and IV of the test, you may need to approximate the credit you would receive for a solution that is not completely correct. Then locate in the adjacent column to the right the scale score that corresponds to your raw score. The scale score is your final Algebra I Regents Examination score.

Regents Examination in Algebra I—August 2017
Chart for Converting Total Test Raw Scores to Final
Examination Scores (Scaled Scores)

Raw Score	Scale Score	Performance Level	Raw Score	Scale Score	Performance Level	Raw Score	Scale Score	Performance Level
86	100	5	57	81	4	28	66	3
85	98	5	56	81	4	27	65	3
84	97	5	55	81	4	26	64	2
83	96	5	54	80	4	25	63	2
82	94	5	53	80	4	24	62	2
81	93	5	52	80	4	23	60	2
80	92	5	51	80	3	22	59	2
79	91	5	50	79	3	21	57	2
78	91	5	49	79	3	20	56	2
77	90	5	48	79	3	19	55	1
76	89	5	47	78	3	18	52	1
75	88	5	46	78	3	17	50	1
74	88	5	45	78	3	16	49	1
73	87	5	44	77	3	15	47	1
72	87	5	43	77	3	14	44	1
71	86	5	42	77	3	13	42	1
70	86	5	41	76	3	12	40	1
69	86	5	40	76	3	11	37	1
68	85	5	39	75	3	10	35	1
67	84	4	38	74	3	9	32	1
66	84	4	37	74	3	8	29	1
65	84	4	36	73	3	7	26	1
64	83	4	35	73	3	6	23	1
63	83	4	34	72	3	5	19	1
62	83	4	33	71	3	4	16	1
61	82	4	32	70	3	3	12	1
60	82	4	31	69	3	2	8	1
59	82	4	30	68	3	1	4	1
58	82	4	29	67	3	0	0	1

Examination
June 2018
Algebra I

HIGH SCHOOL MATH REFERENCE SHEET

Conversions

1 inch = 2.54 centimeters

1 meter = 39.37 inches

1 mile = 5280 feet

1 mile = 1760 yards

1 mile = 1.609 kilometers

1 kilometer = 0.62 mile

1 pound = 16 ounces

1 pound = 0.454 kilogram

1 kilogram = 2.2 pounds

1 ton = 2000 pounds

1 cup = 8 fluid ounces

1 pint = 2 cups

1 quart = 2 pints

1 gallon = 4 quarts

1 gallon = 3.785 liters

1 liter = 0.264 gallon

1 liter = 1000 cubic centimeters

Formulas

Triangle	$A = \frac{1}{2}bh$
Parallelogram	$A = bh$
Circle	$A = \pi r^2$
Circle	$C = \pi d$ or $C = 2\pi r$

Formulas (continued)

General Prisms	$V = Bh$
Cylinder	$V = \pi r^2 h$
Sphere	$V = \frac{4}{3}\pi r^3$
Cone	$V = \frac{1}{3}\pi r^2 h$
Pyramid	$V = \frac{1}{3}Bh$
Pythagorean Theorem	$a^2 + b^2 = c^2$
Quadratic Formula	$x = \dfrac{-b \pm \sqrt{b^2 - 4ac}}{2a}$
Arithmetic Sequence	$a_n = a_1 + (n-1)d$
Geometric Sequence	$a_n = a_1 r^{n-1}$
Geometric Series	$S_n = \dfrac{a_1 - a_1 r^n}{1 - r}$ where $r \neq 1$
Radians	1 radian $= \dfrac{180}{\pi}$ degrees
Degrees	1 degree $= \dfrac{\pi}{180}$ radians
Exponential Growth/Decay	$A = A_0 e^{k(t-t_0)} + B_0$

PART I

Answer all 24 questions in this part. Each correct answer will receive 2 credits. No partial credit will be allowed. For each statement or question, write in the space provided the numeral preceding the word or expression that best completes the statement or answers the question.
[48 credits]

1 The solution to $4p + 2 < 2(p + 5)$ is

   (1) $p > -6$ (3) $p > 4$

   (2) $p < -6$ (4) $p < 4$ 1 _____

2 If $k(x) = 2x^2 - 3\sqrt{x}$, then $k(9)$ is

   (1) 315   (3) 159

   (2) 307   (4) 153   2 _____

3 The expression $3(x^2 + 2x - 3) - 4(4x^2 - 7x + 5)$ is equivalent to

   (1) $-13x - 22x + 11$ (3) $19x^2 - 22x + 11$

   (2) $-13x^2 + 34x - 29$ (4) $19x^2 + 34x - 29$ 3 _____

4 The zeros of the function $p(x) = x^2 - 2x - 24$ are

   (1) -8 and 3   (3) -4 and 6

   (2) -6 and 4   (4) -3 and 8   4 _____

5 The box plot below summarizes the data for the average monthly high temperatures in degrees Fahrenheit for Orlando, Florida.

The third quartile is

   (1) 92   (3) 83

   (2) 90   (4) 71   5 _____

6 Joy wants to buy strawberries and raspberries to bring to a party. Strawberries cost \$1.60 per pound and raspberries cost \$1.75 per pound. If she only has \$10 to spend on berries, which inequality represents the situation where she buys x pounds of strawberries and y pounds of raspberries?

   (1) $1.60x + 1.75y \leq 10$ (3) $1.75x + 1.60y \leq 10$

   (2) $1.60x + 1.75y \geq 10$ (4) $1.75x + 1.60y \geq 10$ 6 _____

7 On the main floor of the Kodak Hall at the Eastman Theater, the number of seats per row increases at a constant rate. Steven counts 31 seats in row 3 and 37 seats in row 6. How many seats are there in row 20?

   (1) 65   (3) 69

   (2) 67   (4) 71   7 _____

8 Which ordered pair below is *not* a solution to $f(x) = x^2 - 3x + 4$?

   (1) $(0, 4)$ (3) $(5, 14)$

   (2) $(1.5, 1.75)$ (4) $(-1, 6)$ 8 _____

9 Students were asked to name their favorite sport from a list of basketball, soccer, or tennis. The results are shown in the table below.

	Basketball	Soccer	Tennis
Girls	42	58	20
Boys	84	41	5

What percentage of the students chose soccer as their favorite sport?

   (1) 39.6%   (3) 50.4%

   (2) 41.4%   (4) 58.6%   9 _____

10 The trinomial $x^2 - 14x + 49$ can be expressed as

   (1) $(x - 7)^2$ (3) $(x - 7)(x + 7)$

   (2) $(x + 7)^2$ (4) $(x - 7)(x + 2)$ 10 _____

11 A function is defined as $\{(0, 1), (2, 3), (5, 8), (7, 2)\}$. Isaac is asked to create one more ordered pair for the function. Which ordered pair can he add to the set to keep it a function?

(1) $(0, 2)$ (3) $(7, 0)$

(2) $(5, 3)$ (4) $(1, 3)$ 11 ____

12 The quadratic equation $x^2 - 6x = 12$ is rewritten in the form $(x + p)^2 = q$, where q is a constant. What is the value of p?

(1) -12 (3) -3

(2) -9 (4) 9 12 ____

13 Which of the quadratic functions below has the *smallest* minimum value?

(1) $h(x) = x^2 + 2x - 6$ (3) $k(x) = (x + 5)(x + 2)$

(2)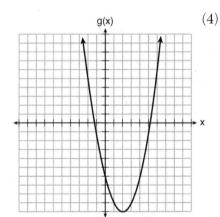

(4)

x	$f(x)$
-1	-2
0	-5
1	-6
2	-5
3	-2

13 ____

14 Which situation is *not* a linear function?

(1) A gym charges a membership fee of $10.00 down and $10.00 per month.

(2) A cab company charges $2.50 initially and $3.00 per mile.

(3) A restaurant employee earns $12.50 per hour.

(4) A $12,000 car depreciates 15% per year.   14 ____

15 The Utica Boilermaker is a 15-kilometer road race. Sara is signed
    up to run this race and has done the following training runs:

> I. 10 miles

> II. 44,880 feet

> III. 15,560 yards

Which run(s) are at least 15 kilometers?

(1) I, only   (3) I and III

(2) II, only   (4) II and III   15 _____

16 If $f(x) = x^2 + 2$, which interval describes the range of this
    function?

(1) $(-\infty, \infty)$ (3) $[2, \infty)$

(2) $[0, \infty)$ (4) $(-\infty, 2]$ 16 _____

17 The amount Mike gets paid weekly can be represented by the
    expression $2.50a + 290$, where a is the number of cell phone
    accessories he sells that week. What is the constant term in this
    expression and what does it represent?

(1) $2.50a$, the amount he is guaranteed to be paid each week

(2) $2.50a$, the amount he earns when he sells a accessories

(3) 290, the amount he is guaranteed to be paid each week

(4) 290, the amount he earns when he sells a accessories   17 _____

18 A cubic function is graphed on the set of axes below.

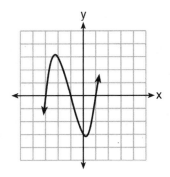

Which function could represent this graph?

(1) $f(x) = (x - 3)(x - 1)(x + 1)$

(2) $g(x) = (x + 3)(x + 1)(x - 1)$

(3) $h(x) = (x - 3)(x - 1)(x + 3)$

(4) $k(x) = (x + 3)(x + 1)(x - 3)$ 18 _____

19 Mrs. Allard asked her students to identify which of the polynomials below are in standard form and explain why.

$$\text{I. } 15x^4 - 6x + 3x^2 - 1$$

$$\text{II. } 12x^3 + 8x + 4$$

$$\text{III. } 2x^5 + 8x^2 + 10x$$

Which student's response is correct?

(1) Tyler said I and II because the coefficients are decreasing.

(2) Susan said only II because all the numbers are decreasing.

(3) Fred said II and III because the exponents are decreasing.

(4) Alyssa said II and III because they each have three terms.   19 _____

20 Which graph does *not* represent a function that is always increasing over the entire interval $-2 < x < 2$?

(1)

(3)

(2)

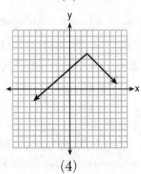

(4)   20 _____

21 At an ice cream shop, the profit, $P(c)$, is modeled by the function $P(c) = 0.87c$, where c represents the number of ice cream cones sold. An appropriate domain for this function is

(1) an integer ≤ 0 (3) a rational number ≤ 0

(2) an integer ≥ 0 (4) a rational number ≥ 0 21 _____

22 How many real-number solutions does $4x^2 + 2x + 5 = 0$ have?

(1) one   (3) zero

(2) two   (4) infinitely many   22 _____

23 Students were asked to write a formula for the length of a rectangle by using the formula for its perimeter, $p = 2\ell + 2w$. Three of their responses are shown below.

$$\text{I.} \quad \ell = \frac{1}{2}p - w$$

$$\text{II.} \quad \ell = \frac{1}{2}(p - 2w)$$

$$\text{III.} \quad \ell = \frac{p - 2w}{2}$$

Which responses are correct?

(1) I and II, only   (3) I and III, only

(2) II and III, only   (4) I, II, and III   23 _____

24 If $a_n = n(a_{n-1})$ and $a_1 = 1$, what is the value of a_5?

(1) 5   (3) 120

(2) 20   (4) 720   24 _____

PART II

Answer all 8 questions in this part. Each correct answer will receive 2 credits. Clearly indicate the necessary steps, including appropriate formula substitutions, diagrams, graphs, charts, etc. For all questions in this part, a correct numerical answer with no work shown will receive only 1 credit.   [16 credits]

25   Graph $f(x) = \sqrt{x + 2}$ over the domain $-2 \leq x \leq 7$.

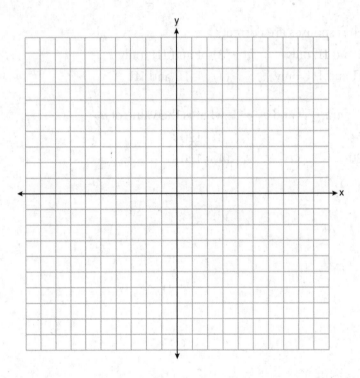

26 Caleb claims that the ordered pairs shown in the table below are from a nonlinear function.

x	$f(x)$
0	2
1	4
2	8
3	16

State if Caleb is correct. Explain your reasoning.

27 Solve for x to the *nearest tenth*: $x^2 + x - 5 = 0$.

28 The graph of the function $p(x)$ is represented below. On the same set of axes, sketch the function $p(x + 2)$.

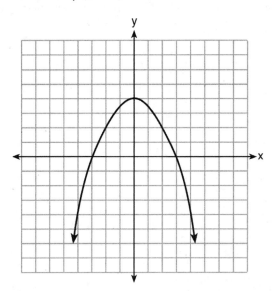

29 When an apple is dropped from a tower 256 feet high, the function $h(t) = -16t^2 + 256$ models the height of the apple, in feet, after t seconds. Determine, algebraically, the number of seconds it takes the apple to hit the ground.

30 Solve the equation below algebraically for the exact value of x.

$$6 - \frac{2}{3}(x + 5) = 4x$$

31 Is the product of $\sqrt{16}$ and $\frac{4}{7}$ rational or irrational?

Explain your reasoning.

32 On the set of axes below, graph the piecewise function:

$$f(x) = \begin{cases} -\dfrac{1}{2}x, & x < 2 \\ x, & x \geq 2 \end{cases}$$

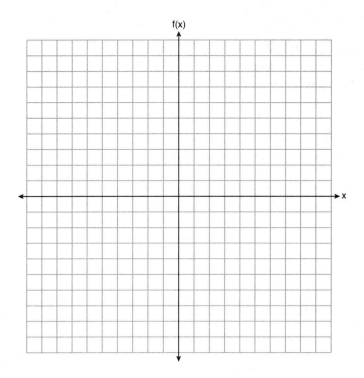

PART III

Answer all 4 questions in this part. Each correct answer will receive 4 credits. Clearly indicate the necessary steps, including appropriate formula substitutions, diagrams, graphs, charts, etc. For all questions in this part, a correct numerical answer with no work shown will receive only 1 credit. [16 credits]

33 A population of rabbits in a lab, $p(x)$, can be modeled by the function $p(x) = 20(1.014)^x$, where x represents the number of days since the population was first counted.

Explain what 20 and 1.014 represent in the context of the problem.

Determine, to the *nearest tenth*, the average rate of change from day 50 to day 100.

34 There are two parking garages in Beacon Falls. Garage *A* charges $7.00 to park for the first 2 hours, and each additional hour costs $3.00. Garage *B* charges $3.25 per hour to park.

When a person parks for at least 2 hours, write equations to model the cost of parking for a total of *x* hours in Garage *A* and Garage *B*.

Determine algebraically the number of hours when the cost of parking at both garages will be the same.

35 On the set of axes below, graph the following system of inequalities:

$$2y + 3x \leq 14$$
$$4x - y < 2$$

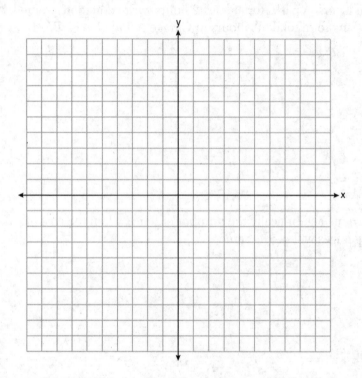

Determine if the point $(1, 2)$ is in the solution set. Explain your answer.

36 The percentage of students scoring 85 or better on a mathematics final exam and an English final exam during a recent school year for seven schools is shown in the table below.

Percentage of Students Scoring 85 or Better	
Mathematics, x	English, y
27	46
12	28
13	45
10	34
30	56
45	67
20	42

Write the linear regression equation for these data, rounding all values to the *nearest hundredth*.

State the correlation coefficient of the linear regression equation, to the *nearest hundredth*. Explain the meaning of this value in the context of these data.

PART IV

**Answer the question in this part. A correct answer will receive 6 credits.
Clearly indicate the necessary steps, including appropriate formula substitutions, diagrams, graphs, charts, etc. A correct numerical answer
with no work shown will receive only 1 credit.**   [6 credits]

37 Dylan has a bank that sorts coins as they are dropped into it. A panel on the
    front displays the total number of coins inside as well as the total value of
    these coins. The panel shows 90 coins with a value of $17.55 inside of the
    bank.

    If Dylan only collects dimes and quarters, write a system of equations in two
    variables or an equation in one variable that could be used to model this
    situation.

    Using your equation or system of equations, algebraically determine the
    number of quarters Dylan has in his bank.

Question 37 is continued on the next page.

Question 37 continued

Dylan's mom told him that she would replace each one of his dimes with a quarter. If he uses all of his coins, determine if Dylan would then have enough money to buy a game priced at $20.98 if he must also pay an 8% sales tax. Justify your answer.

Answers
June 2018
Algebra I

Answer Key

PART I

1. (4)	**5.** (2)	**9.** (1)	**13.** (2)	**17.** (3)	**21.** (2)
2. (4)	**6.** (1)	**10.** (1)	**14.** (4)	**18.** (2)	**22.** (3)
3. (2)	**7.** (1)	**11.** (4)	**15.** (1)	**19.** (3)	**23.** (4)
4. (3)	**8.** (4)	**12.** (3)	**16.** (3)	**20.** (3)	**24.** (3)

PART II

25.

26. Caleb is correct. It is a nonlinear function.

27. $x = 1.8$ and $x = -2.8$

28. See graph on page 327.

29. 4 seconds

30. $\frac{4}{7}$

31. The product is rational.

32. See graphs on pages 329 and 330.

PART III

33. 20 is the starting population. 1.014 is the growth factor. The average rate of change is 0.8.

34. For Garage A, $A = 7 + 3(x - 2)$. For Garage B, $B = 3.25x$. Both garages cost the same when you park for 4 hours.

35. $(1, 2)$ is not in the solution set.

36. $y = 0.96x + 23.95$ The correlation coefficient is 0.92, which is a strong positive correlation.

PART IV

37. $0.10d + 0.25q = 17.55$
$$d + q = 90$$

Dylan has 57 quarters. No, he would not be able to buy the game.

In **Parts II–IV**, you are required to show how you arrived at your answers. For sample methods of solutions, see the *Answers Explained* section.

Answers and Explanations

PART I

1. Solving an inequality is very similar to solving an equality. Use the properties of algebra to isolate the variable.

   First distribute the 2 on the right-hand side:

   $$4p + 2 < 2(p + 5)$$
   $$4p + 2 < 2p + 10$$

   Then eliminate the constant term from the left-hand side:

   $$4p + 2 < 2p + 10$$
   $$-2 = -2$$
   $$4p < 2p + 8$$

   Next eliminate the variables from the right-hand side:

   $$4p < 2p + 8$$
   $$-2p = -2p$$
   $$2p < 8$$

   Finally, eliminate the coefficient by dividing both sides of the inequality by 2. If the coefficient is negative, the direction of the inequality would need to be reversed. In this case, the coefficient is positive. So do not reverse the inequality:

   $$\frac{2p}{2} < \frac{8}{2}$$
   $$p < 4$$

   The correct choice is (**4**).

2. Substitute 9 for all occurrences of x in the function:

$$k(9) = 2(9)^2 - 3\sqrt{9}$$
$$= 2(81) - 3(3)$$
$$= 162 - 9$$
$$= 153$$

The correct choice is (**4**).

3. Distribute the 3 to the terms in the first parentheses:

$$3x^2 + 6x - 9 - 4(4x^2 - 7x + 5)$$

Then distribute the -4 to the terms in the second parentheses. Be sure to change the signs since you are multiplying by a negative:

$$3x^2 + 6x - 9 - 16x^2 + 28x - 20$$

Now combine like terms:

$$3x^2 - 16x^2 + 6x + 28x - 9 - 20 = -13x^2 + 34x - 29$$

The correct choice is (**2**).

4. The zeros of a function are the numbers that evaluate to zero when you plug them into the function. To find the zeros, solve the equation $x^2 - 2x - 24 = 0$. There are several ways to solve this equation. The simplest is by factoring:

$$x^2 - 2x - 24 = 0$$
$$(x - 6)(x + 4) = 0$$
$$x - 6 = 0 \text{ or } x + 4 = 0$$
$$x = 6 \text{ or } x = -4$$

Another way to find the zeros of a function is to graph it on a graphing calculator and find the x-intercepts of the graph. This can be done with the zero function on the calculator.

For the TI-84:

For the TI-Nspire:

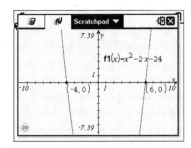

The correct choice is **(3)**.

5. In a box plot, the five values marked with vertical lines are the minimum, the first quartile, the median, the third quartile, and the maximum. For this box plot, the minimum is about 71, the first quartile is 75, the median is about 83, the third quartile is 90, and the maximum is about 92.

   The correct choice is **(2)**.

6. Since 1 pound of strawberries costs $1.60, x pounds of strawberries cost $1.60x$. Since 1 pound of raspberries costs 1.75, y pounds of raspberries cost $1.75y$. Together, the cost of x pounds of strawberries and y pounds of raspberries is $1.60x + 1.75y$. Since Joy has only $10 to spend, the total cost must be less than or equal to 10. So the inequality that represents the situation is $1.60x + 1.75y \le 10$.

   The correct choice is **(1)**.

7. One way to solve this is to make a list with the given information:

Row 1.

Row 2.

Row 3. 31 seats

Row 4.

Row 5.

Row 6. 37 seats

The seats increase at a constant rate. Since the number of seats increases by 6 when the number of rows increases by 3, the increase in seats for each new row must be $6 \div 3 = 2$. Create a table to show the number of seats in each row:

1. 27	5. 35	9. 43	13. 51	17. 59
2. 29	6. 37	10. 45	14. 53	18. 61
3. 31	7. 39	11. 47	15. 55	19. 63
4. 33	8. 41	12. 49	16. 57	20. 65

A quicker way to solve this problem is to use the arithmetic sequence formula from the reference sheet, $a_n = a_1 + (n - 1)d$. Remember that a_1 is the first term of the sequence, n is the term you want to find, and d is the difference from one term to the next.

For this question, $a_1 = 27$, $d = 2$, and $n = 20$. Plug the values into the formula:

$$
\begin{aligned}
a_{20} &= 27 + (20 - 1)2 \\
&= 27 + 19 \cdot 2 \\
&= 27 + 38 \\
&= 65
\end{aligned}
$$

A third way to solve this problem is to treat x as the row number and y as the number of seats in row x. Then find the equation of the line in slope-intercept form that passes through $(3, 31)$ and $(6, 37)$:

$$m = \frac{37 - 31}{6 - 3}$$
$$= \frac{6}{3}$$
$$= 2$$
$$y = mx + b$$
$$37 = 2 \cdot 6 + b$$
$$37 = 12 + b$$
$$-12 = -12$$
$$25 = b$$
$$y = 2x + 25$$

Substitute 20 for x:

$$y = 2 \cdot 20 + 25$$
$$= 40 + 25$$
$$= 65$$

The correct choice is **(1)**.

8. To check if an ordered pair is a solution, plug in the x-coordinate of the ordered pair and see if the function evaluates to the corresponding y-coordinate.

- Testing choice (1):

$$f(0) = 0^2 - 3 \cdot 0 + 4 \overset{\checkmark}{=} 4$$

- Testing choice (2):

$$f(1.5) = 1.5^2 - 3 \cdot 1.5 + 4 \overset{\checkmark}{=} 1.75$$

- Testing choice (3):

$$f(5) = 5^2 - 3 \cdot 5 + 4 \overset{\checkmark}{=} 14$$

- Testing choice (4):

$$f(-1) = (-1)^2 - 3 \cdot (-1) + 4 = 1 + 3 + 4 = 8 \neq 6$$

The correct choice is **(4)**.

9. To calculate the percentage of students who chose soccer as their favorite sport, divide the number of students who chose soccer by the total number of students. There were 58 boys who chose soccer and 41 girls who chose soccer. So $58 + 41 = 99$ students chose soccer. The total number of students who answered the question was

$$42 + 58 + 20 + 84 + 41 + 5 = 250$$

The percentage of students who chose soccer was

$$\frac{99}{250} \approx 39.6\%$$

The correct choice is **(1)**.

10. You can factor an expression of the form $x^2 + bx + c$ into the product $(x + r_1)$ $(x + r_2)$ by finding two numbers whose sum is b and whose product is c. In this problem, the two numbers whose sum is -14 and whose product is 49 are -7 and -7. So the expression factors into $(x - 7)(x - 7) = (x - 7)^2$.

Another way to solve this is to recognize that in an expression like $x^2 + bx + c$ when the c-term is equal to the square of half the b-term, the expression is a perfect square trinomial. That expression can be factored into $\left(x + \frac{b}{2}\right)^2$. In this example, $\left(\frac{-14}{2}\right)^2 = 49$. So the expression factors into $\left(x + \frac{-14}{2}\right)^2 = (x - 7)^2$.

Since this is a multiple-choice question, a third way this can be solved is to test each of the four choices to see which one becomes $x^2 - 14x + 49$ when multiplied.

- Testing choice (1):

$$(x - 7)^2 = (x - 7)(x - 7)$$
$$= x^2 - 7x - 7x + 49$$
$$= x^2 - 14x + 49$$

- Testing choice (2):

$$(x + 7)^2 = (x + 7)(x + 7)$$
$$= x^2 + 7x + 7x + 49$$
$$= x^2 + 14x + 49$$

- Testing choice (3):

$$(x - 7)(x + 7) = x^2 + 7x - 7x - 49$$
$$= x^2 - 49$$

- Testing choice (4):

$$(x - 7)(x + 2) = x^2 + 2x - 7x - 14$$
$$= x^2 - 5x - 14$$

The correct choice is **(1)**.

11. When a function is described as a set of ordered pairs, none of the x-coordinates can repeat.

- Testing choice (1): $(0, 2)$ cannot be part of the function since the ordered pair $(0, 1)$ already has 0 as an x-coordinate.
- Testing choice (2): $(5, 3)$ cannot be part of the function since the ordered pair $(5, 8)$ already has 5 as an x-coordinate.
- Testing choice (3): $(7, 0)$ cannot be part of the function since the ordered pair $(7, 2)$ already has 7 as an x-coordinate.
- Testing choice (4): $(1, 3)$ can be part of the function since no other ordered pairs in the set have an x-coordinate of 1. It does not matter that the ordered pair $(2, 3)$ already has a y-coordinate of 3.

The correct choice is **(4)**.

12. This question requires that you do the first two steps of solving a quadratic equation by completing the square process.

Step 1: The first step of solving an equation of the form $x^2 + bx = c$ by completing the square is to add $\left(\dfrac{b}{2}\right)^2$ to both sides of the equation.

For this example:

$$x^2 - 6x + \left(\dfrac{-6}{2}\right)^2 = 12 + \left(\dfrac{-6}{2}\right)^2$$
$$x^2 - 6x + 9 = 12 + 9$$
$$x^2 - 6x + 9 = 21$$

Step 2: Factor the left side of the equation:

$$(x - 3)(x - 3) = 21$$
$$(x - 3)^2 = 21$$

The completing the square process is not finished. Since this question just asks for the value of p, though, there is no more work to do. If $(x + p) = (x - 3)$, then p must be -3.

The correct choice is **(3)**.

13. The quadratic function with the smallest minimum value is the one whose vertex has the smallest y-coordinate.

- Testing choice (1): Use the graphing calculator to find the vertex of the parabola. The minimum is at $(-1, -7)$.

For the TI-84:

For the TI-Nspire:

- Testing choice (2): As shown on the graph, the vertex of the parabola is at $(2, -10)$.
- Testing choice (3): Use the graphing calculator to find the vertex of the parabola. The minimum is at $(-3.5, -2.25)$.
- Testing choice (4): When you graph the five given points, they resemble a parabola with a vertex at $(1, -6)$.

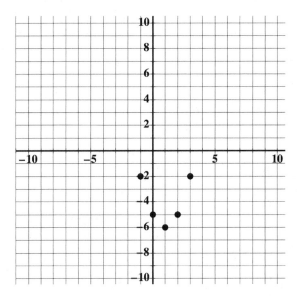

Since -10 is the minimum of all the y-coordinates of the four vertices, it is the smallest minimum value.

The correct choice is (2).

14. One way to answer this question is to create an equation to model each scenario and determine which equation is not in the form $y = mx + b$.

- Testing choice (1): The equation is $y = 10x + 10$.
- Testing choice (2): The equation is $y = 3x + 2.5$.
- Testing choice (3): The equation is $y = 12.5x + 0$.
- Testing choice (4): The equation is $y = 12{,}000 \cdot 0.85^x$.

Choices (1), (2), and (3) are linear functions, while choice (4) is an exponential function.

Another way to answer this question is to make a sketch of each scenario and see which of them does not form a line.

- Testing choice (1):

- Testing choice (2):

- Testing choice (3):

- Testing choice (4):

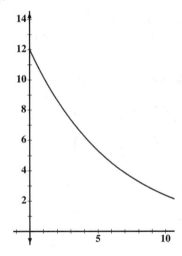

Only choice (4) does not form a line.

The correct choice is **(4)**.

15. Use the reference sheet to convert the measurements in each of the three runs into kilometers.

- Run I:

$$1 \text{ mile} = 1.609 \text{ kilometers}$$

$$10 \text{ miles}\left(\frac{1.609 \text{ kilometers}}{1 \text{ mile}}\right) = 16.09 \text{ kilometers}$$

$$16.09 \text{ kilometers} \geq 15 \text{ kilometers}$$

- Run II:

$$5280 \text{ feet} = 1 \text{ mile}$$

$$44,880 \text{ feet}\left(\frac{1 \text{ mile}}{5280 \text{ feet}}\right) = 8.5 \text{ miles}$$

$$1 \text{ mile} = 1.609 \text{ kilometers}$$

$$8.5 \text{ miles}\left(\frac{1.609 \text{ kilometers}}{1 \text{ mile}}\right) = 13.6765 \text{ kilometers} \approx 13.7 \text{ kilometers}$$

$$13.7 \text{ kilometers} < 15 \text{ kilometers}$$

- Run III:

$$1760 \text{ yards} = 1 \text{ mile}$$

$$15,560 \text{ yards}\left(\frac{1 \text{ mile}}{1760 \text{ yards}}\right) = 8.8409 \text{ miles} \approx 8.84 \text{ miles}$$

$$1 \text{ mile} = 1.609 \text{ kilometers}$$

$$8.84 \text{ miles}\left(\frac{1.609 \text{ kilometers}}{1 \text{ mile}}\right) = 14.22356 \text{ kilometers} \approx 14.2 \text{ kilometers}$$

$$14.2 \text{ kilometers} < 15 \text{ kilometers}$$

The correct choice is (**1**).

16. The range of a function is the set of all the different y-coordinates of the graph of that function. You can graph this function on the graphing calculator.

For the TI-84:

For the TI-Nspire:

The minimum point has a y-coordinate of 2, and there is no maximum point. So the range is $y \geq 2$. In interval notation, this is written as $[2, \infty)$. The square bracket indicates that the interval includes the endpoint of 2. For the ∞, you always use a rounded parenthesis.

The correct choice is **(3)**.

17. In an expression in the form $mx + b$, the b is the constant term, the m is the coefficient, the x is the variable, and the mx is the variable part. In the expression $2.50a + 290$, the number 290 is the constant term.

When an expression like this is used to model a real-world scenario, you can interpret the meaning of the coefficient and of the constant by plugging in 0 and 1 for the variable.

If you plug in 0 for a, you get $2.50 \cdot 0 + 290 = 290$. So 290 is the amount Mike is guaranteed to be paid even if he sells 0 cell phone accessories.

If you plug in 1 for a, you get $2.50 \cdot 1 + 290 = 292.50$. So 2.50 must be the amount Mike earns for each accessory he sells. Therefore $2.50a$ is the amount Mike earns when he sells a accessories.

The correct choice is **(3)**.

18. If the graph of a polynomial function has an x-intercept at $(a, 0)$, the function will have a factor of $(x - a)$. Since the graph of this function has x-intercepts at $(-3, 0)$, $(-1, 0)$, and $(1, 0)$, the function must have factors of $(x - (-3)) = (x + 3)$, $(x - (-1)) = (x + 1)$, and $(x - (+1)) = (x - 1)$.

Another way to answer this is to graph the four choices to see which most resembles the given graph. On the graphing calculator, choice (2) looks most like the given graph.

For the TI-84:

For the TI-Nspire:

The correct choice is **(2)**.

19. A polynomial is in standard form if the exponents are decreasing from left to right. Whether the coefficients are decreasing, increasing, or fluctuating up and down makes no difference.

- For option I, the exponents go from 4 to 1 to 2 to 0, which is not always decreasing.
- For option II, the exponents decrease from 3 to 1 to 0.
- For option III, the exponents decrease from 5 to 2 to 1.

Both options II and III are in standard form because their exponents decrease in value. Both choices (3) and (4) identify which of the Roman numerals are in standard form. However, choice (4) is incorrect since the number of terms has nothing to do with whether or not a polynomial is in standard form.

The correct choice is **(3)**.

20. A function is increasing over an interval if the graph is always going "up" as you move from left to right.

    - Testing choice (1): Between $x = -2$ and $x = 2$, the function is always increasing.
    - Testing choice (2): Between $x = -2$ and $x = 2$, the function is always increasing. There is a "jump" at $x = -2$. So it isn't accurate to say that this function is always increasing on all intervals. However between $x = -2$ and $x = 2$, it is increasing.
    - Testing choice (3): Between $x = -2$ and approximately $x = 0$, the function is increasing. From approximately $x = 0$ to $x = 2$, the function is decreasing.
    - Testing choice (4): Even though this function starts to decrease at $x = 3$, it is increasing from $x = -2$ to $x = 2$.

    The correct choice is **(3)**.

21. Since you cannot generally buy a fraction of an ice cream cone, the domain must be integers. Since you cannot buy a negative number of ice cream cones, the domain must be a nonnegative integer (including 0). If you were permitted to buy fractions of an ice cream cone, the answer would be choice (4).

    The correct choice is **(2)**.

22. The quickest way to solve this question is to graph the parabola $y = 4x^2 + 2x + 5$ and count how many x-intercepts it has.

For the TI-84:

For the TI-Nspire:

If you want to use algebra instead, calculate $b^2 - 4ac$. If $b^2 - 4ac = 0$, the equation has one real solution. If $b^2 - 4ac > 0$, the equation has two real solutions. If $b^2 - 4ac < 0$, the equation has no real solutions.

In this example, $b^2 - 4ac = 2^2 - 4 \cdot 4 \cdot 5 = 4 - 80 = -76 < 0$. So the equation has no real solutions.

The correct choice is (3).

23. Isolate the l in the equation.

Step 1: Subtract $2w$ from both sides of the equation:

$$p = 2l + 2w$$
$$-2w = -2w$$
$$p - 2w = 2l$$

Step 2: Divide both sides by 2:

$$\frac{p - 2w}{2} = \frac{2l}{2}$$
$$\frac{p - 2w}{2} = l$$
$$l = \frac{p - 2w}{2}$$

This is option III.

Since dividing by 2 is the same thing as multiplying by $\frac{1}{2}$, the equation can also be written as:

$$\frac{1}{2}(p - 2w) = l$$
$$l = \frac{1}{2}(p - 2w)$$

This is option II.

If you distribute the $\frac{1}{2}$ through the right side of this equation you get:

$$l = \frac{1}{2}p - \frac{1}{2} \cdot 2w$$
$$l = \frac{1}{2}p - w$$

This is option I.

The correct choice is **(4)**.

24. To calculate a_5 with this recursive definition of the function, you first have to calculate a_2, a_3, and a_4.

For $n = 2$:

$$
\begin{aligned}
a_2 &= 2\left(a_{2-1}\right) \\
&= 2\left(a_1\right) \\
&= 2 \cdot 1 \\
&= 2
\end{aligned}
$$

For $n = 3$:

$$
\begin{aligned}
a_3 &= 3\left(a_{3-1}\right) \\
&= 3\left(a_2\right) \\
&= 3 \cdot 2 \\
&= 6
\end{aligned}
$$

For $n = 4$:

$$
\begin{aligned}
a_4 &= 4\left(a_{4-1}\right) \\
&= 4\left(a_3\right) \\
&= 4 \cdot 6 \\
&= 24
\end{aligned}
$$

For $n = 5$:

$$
\begin{aligned}
a_5 &= 5\left(a_{5-1}\right) \\
&= 5\left(a_4\right) \\
&= 5 \cdot 24 \\
&= 120
\end{aligned}
$$

The correct choice is (**3**).

PART II

25. First make a chart for all integer values from -2 to 7:

x	$f(x)$
-2	$\sqrt{-2+2} = \sqrt{0} = 0$
-1	$\sqrt{-1+2} = \sqrt{1} = 1$
0	$\sqrt{0+2} = \sqrt{2} \approx 1.4$
1	$\sqrt{1+2} = \sqrt{3} \approx 1.7$
2	$\sqrt{2+2} = \sqrt{4} = 2$
3	$\sqrt{3+2} = \sqrt{5} \approx 2.2$
4	$\sqrt{4+2} = \sqrt{6} \approx 2.4$
5	$\sqrt{5+2} = \sqrt{7} \approx 2.6$
6	$\sqrt{6+2} = \sqrt{8} \approx 2.8$
7	$\sqrt{7+2} = \sqrt{9} = 3$

Plot the 10 points on the graph, and join the points with a curve.

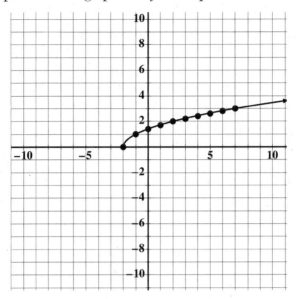

26. The graph of a nonlinear function is not a line. When the four ordered pairs are plotted, the graph looks like this:

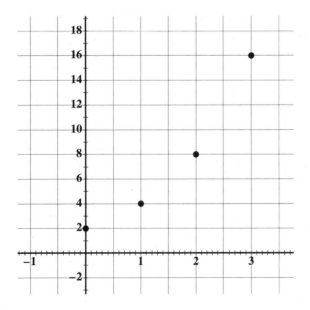

Since there is no line that passes through all four of these points, Caleb is correct. The ordered pairs are from a nonlinear function.

Even without graphing, there is a way to tell that the ordered pairs are from a nonlinear function. In a linear function every time x increases by 1, $f(x)$ increases by a specific amount. In this example when x increases from 0 to 1, $f(x)$ increases by 2. When x increases from 1 to 2, though, $f(x)$ increases by 4. This can happen only in a nonlinear function.

Caleb is correct. It is a nonlinear function.

27. Use the quadratic formula from the reference sheet, $x = \dfrac{-b \pm \sqrt{b^2 - 4ac}}{2a}$.

For this question, $a = 1, b = 1$, and $c = -5$.

$$x = \frac{-1 \pm \sqrt{1^2 - 4 \cdot 1 \cdot (-5)}}{2 \cdot 1}$$

$$= \frac{-1 \pm \sqrt{1 + 20}}{2}$$

$$= \frac{-1 \pm \sqrt{21}}{2}$$

$$\frac{-1 + \sqrt{21}}{2} \approx 1.79 \approx 1.8$$

$$\frac{-1 - \sqrt{21}}{2} \approx -2.79 \approx -2.8$$

To the *nearest tenth*, $x = 1.8$ and $x = -2.8$.

28. The graph of $p(x + 2)$ is the graph of $p(x)$ shifted to the *left* by 2 units.

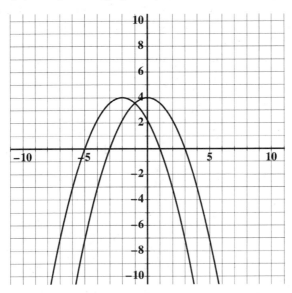

29. When the apple hits the ground, the height of the apple will be 0. To determine algebraically when this will happen, solve the equation $0 = -16t^2 + 256$. The answer must be positive since the number of seconds for the apple to fall cannot be negative:

$$0 = -16t^2 + 256$$
$$+16t^2 = +16t^2$$
$$\frac{16t^2}{16} = \frac{256}{16}$$
$$t^2 = 16$$
$$t = \sqrt{16} = 4$$

The apple takes 4 seconds to hit the ground.

30. Start with the distributive property. Then use the properties of algebra to get the constants on one side of the equals sign and the variables on the other side of the equals sign. To finish solving the equation, isolate the x variable:

$$6 - \frac{2}{3}(x + 5) = 4x$$

$$6 - \frac{2}{3}x - \frac{2}{3} \cdot 5 = 4x$$

$$6 - \frac{2}{3}x - \frac{10}{3} = 4x$$

$$+\frac{2}{3}x = +\frac{2}{3}x$$

$$6 - \frac{10}{3} = 4x + \frac{2}{3}x$$

$$\frac{18}{3} - \frac{10}{3} = \frac{12}{3}x + \frac{2}{3}x$$

$$\frac{8}{3} = \frac{14}{3}x$$

$$\frac{3}{14} \cdot \frac{8}{3} = \frac{3}{14} \cdot \frac{14}{3}x$$

$$\frac{8}{14} = x$$

$$\frac{4}{7} = x$$

The exact value of x is $\frac{4}{7}$.

31. A rational number is one that can be written as a fraction that has an integer in both the numerator and the denominator. Since 16 is a perfect square, $\sqrt{16} = 4$ is a rational number. The product of 4 and $\frac{4}{7}$ is $4 \cdot \frac{4}{7} = \frac{16}{7}$. Since this fraction has an integer in both the numerator and the denominator, it is rational.

The product is rational.

32. Since the boundary between the two pieces is $x = 2$, make a chart with some x-values less than 2 and some x-values greater than 2. For $x = -1$, 0, and 1, use the first equation in the function. For $x = 2, 3, 4$, and 5, use the second equation in the function:

x	$f(x)$
-1	$-\frac{1}{2} \cdot (-1) = \frac{1}{2}$
0	$-\frac{1}{2} \cdot 0 = 0$
1	$-\frac{1}{2} \cdot 1 = -\frac{1}{2}$
2	2
3	3
4	4
5	5

This is what the 7 points look like on the graph:

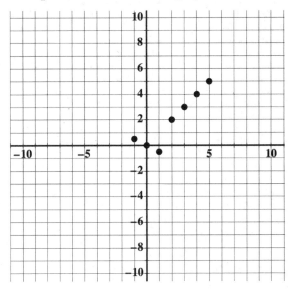

To plot the first piece of the graph, find the boundary point by plugging 2 into the first equation $-\frac{1}{2} \cdot 2 = -1$. This corresponds to the point $(2, -1)$.

Even though $(2, -1)$ is not an actual point on the graph, it is important to plot it on the graph as a small open circle. Make a ray that starts at this open circle and that passes through $\left(1, -\frac{1}{2}\right)$.

To plot the second piece of the the graph, make a ray that starts at $(2, 2)$ and that goes through $(3, 3)$.

The finished graph looks like this:

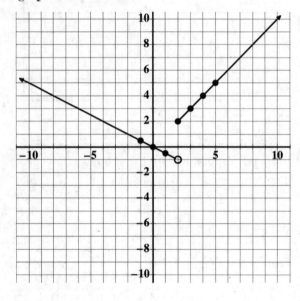

PART III

33. In an exponential equation of the form $y = a \cdot b^x$, a represents the starting amount and b represents the growth factor. In this question, the 20 represents the number of rabbits when the population was first counted. The 1.014 is the growth factor. The growth factor is also 1 plus the growth rate, so this population is increasing by 1.4% a year.

The average rate of change of a function $p(x)$ between $x = a$ and $x = b$ can be calculated with the formula $\dfrac{p(b) - p(a)}{b - a}$. In this question, $a = 50$ and $b = 100$. Calculate the average rate of change:

$$\frac{p(100) - p(50)}{100 - 50} = \frac{20(1.014)^{100} - 20(1.014)^{50}}{100 - 50}$$

$$= \frac{80.3 - 40.1}{100 - 50}$$

$$= \frac{40.2}{50}$$

$$= 0.804 \approx 0.8$$

To the *nearest tenth*, the average rate of change from day 50 to day 100 is 0.8.

34. For Garage A, the cost of parking for x hours (where x is greater than 2) is $A = 7 + 3(x - 2)$. For example, if you park for 6 hours in Garage A, the cost is \$7 for the first 2 hours and then \$3 for the additional 4 (not 6) hours. A common mistake is to write $A = 7 + 3x$ for the cost. However, that equation would make you pay an extra \$3 for each of the first 2 hours.

For Garage B, the cost of parking for x hours is $B = 3.25x$.

The cost of parking at both garages will be the same when $7 + 3(x - 2) = 3.25x$.

Algebraically, this can be solved as follows:

$$7 + 3(x - 2) = 3.25x$$
$$7 + 3x - 6 = 3.25x$$
$$-3x = -3x$$
$$7 - 6 = 3.25x - 3x$$
$$\frac{1}{0.25} = \frac{0.25x}{0.25}$$
$$4 = x$$

Both garages cost the same when you park for 4 hours.

35. Linear inequalities are graphed as a boundary line and all the points on one side of the line are shaded. To graph the inequality $2y + 3x \le 14$, first graph the boundary line. Start by changing the inequality sign into an equals sign. Then rewrite the equation in slope-intercept form $y = mx + b$:

$$2y + 3x \le 14$$
$$2y + 3x = 14$$
$$-3x = -3x$$
$$\frac{2y}{2} = \frac{-3x + 14}{2}$$
$$y = -\frac{3}{2}x + 7$$

Since the inequality has a $\le$ instead of a $<$, graph a solid line.

To determine which side of the line to shade, test if the point $(0, 0)$ makes the inequality true. Substitute $(0, 0)$ into the original inequality $2y + 3x \le 14$:

$$2 \cdot 0 + 3 \cdot 0 \le 14$$
$$0 + 0 \le 14$$
$$0 \le 14$$

Since it is true that 0 is less than or equal to 14, shade the side of the boundary line that contains $(0, 0)$.

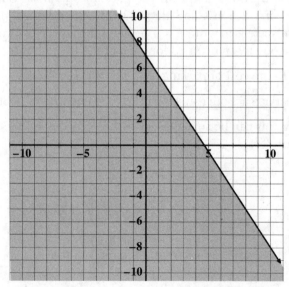

To find the boundary line, change the inequality $4x - y < 2$ into an equation by substituting an equals sign for the inequality sign. Then write the equation in slope-intercept form $y = mx + b$:

$$4x - y < 2$$

$$4x - y = 2$$
$$-4x = -4x$$
$$\frac{-y}{-1} = \frac{-4x + 2}{-1}$$
$$y = 4x - 2$$

Since the inequality has a $<$ instead of a $\leq$, graph a dotted line.

Test to see if $(0, 0)$ is a solution to the second inequality:

$$4 \cdot 0 - 0 < 2$$
$$0 - 0 < 2$$
$$0 < 2$$

Since this is true, shade the side of the line that contains $(0, 0)$.

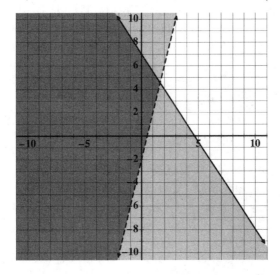

Inequalities can also be graphed using the graphing calculator.

For the TI-84:

For the TI-Nspire:

A point is in the solution set if it is in the double-shaded area. The point $(1, 2)$ is on the dotted line, so it is not part of the double-shaded area and, therefore, is not part of the solution set. If it was in the double-shaded area or even on the solid boundary line, it would be part of the solution set.

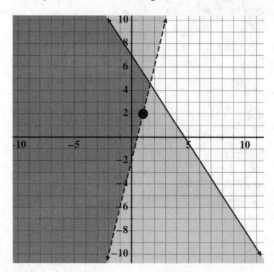

36. Enter the data into the graphing calculator, and use the linear regression function.

For the TI-84:

If your TI-84 calculator is not showing the value of r, which is the correlation coefficient, you have to turn on the diagnostics by pressing [2nd] [0] and scrolling down to DiagnosticOn.

For the TI-Nspire:

To the *nearest hundredth*, the equation is $y = 0.96x + 23.95$.

The correlation coefficient is 0.92, which is very close to $+1$. This means that there is a strong positive correlation. If the correlation coefficient was very close to -1, there would be a strong negative correlation. If the correlation coefficient was close to 0, there would be a weak correlation.

PART IV

37. If d is the number of dimes and q is the number of quarters, the following system of equations can model this situation:

$$0.10d + 0.25q = 17.55$$
$$d + q = 90$$

To solve this system of equations, start by isolating the d in the second equation. Then use the substitution method so the only remaining variable is q, which shows the number of quarters Dylan has in his bank:

$$d = 90 - q$$
$$0.10d + 0.25q = 17.55$$
$$0.10(90 - q) + 0.25q = 17.55$$
$$9 - 0.10q + 0.25q = 17.55$$
$$-9 = -9$$
$$-0.10q + 0.25q = 17.55 - 9$$
$$\frac{0.15q}{0.15} = \frac{8.55}{0.15}$$
$$q = 57$$

Dylan has 57 quarters in his bank.

If each of Dylan's dimes is replaced with a quarter, he would have 90 quarters. They would be worth:

$$90 \times \$0.25 = \$22.50$$

The game costs $20.98 plus 8% of $20.98 for sales tax:

$$\$20.98 + (0.08 \times \$20.98) = \$22.66$$

Including sales tax, the game costs $22.66, which is more than $22.50.

No, Dylan would not be able to buy the game if all his dimes are replaced with quarters.

Topic	Question Numbers	Number of Points	Your Points	Your Percentage
1. Polynomials	3, 12, 19	$2 + 2 + 2 = 6$		
2. Properties of Algebra	30	2		
3. Functions	2, 8, 11, 20	$2 + 2 + 2 + 2 = 8$		
4. Creating and Interpreting Equations	6, 14, 17, 21, 23	$2 + 2 + 2 + 2 + 2 = 10$		
5. Inequalities	1, 35	$2 + 4 = 6$		
6. Sequences and Series	7, 24	$2 + 2 = 4$		
7. Systems of Equations	34, 37	$4 + 6 = 10$		
8. Quadratic Equations and Factoring	4, 10, 22, 27, 29	$2 + 2 + 2 + 2 + 2 = 10$		
9. Regression	36	4		
10. Exponential Equations	26, 33	$2 + 4 = 6$		
11. Graphing	13, 16, 18, 25, 28, 32	$2 + 2 + 2 + 2 + 2 + 2 = 12$		
12. Statistics	5, 9	$2 + 2 = 4$		
13. Number Properties	31	2		
13. Unit Conversions	15	2		

HOW TO CONVERT YOUR RAW SCORE TO YOUR ALGEBRA I REGENTS EXAMINATION SCORE

The accompanying conversion chart must be used to determine your final score on the June 2018 Regents Examination in Algebra I. To find your final exam score, locate in the column labeled "Raw Score" the total number of points you scored out of a possible 86 points. Since partial credit is allowed in Parts II, III, and IV of the test, you may need to approximate the credit you would receive for a solution that is not completely correct. Then locate in the adjacent column to the right the scale score that corresponds to your raw score. The scale score is your final Algebra I Regents Examination score.

Regents Examination in Algebra I—June 2018
Chart for Converting Total Test Raw Scores to Final
Examination Scores (Scaled Scores)

Raw Score	Scale Score	Performance Level	Raw Score	Scale Score	Performance Level	Raw Score	Scale Score	Performance Level
86	100	5	57	81	4	28	67	3
85	99	5	56	81	4	27	66	3
84	98	5	55	81	4	26	65	4
83	97	5	54	81	4	25	63	2
82	96	5	53	80	4	24	62	2
81	94	5	52	80	4	23	60	2
80	93	5	51	80	4	22	59	2
79	93	5	50	80	4	21	57	2
78	92	5	49	79	3	20	56	2
77	91	5	48	79	3	19	55	2
76	90	5	47	79	3	18	52	1
75	89	5	46	78	3	17	51	1
74	89	5	45	78	3	16	49	1
73	88	5	44	78	3	15	47	1
72	87	5	43	77	3	14	44	1
71	87	5	42	77	3	13	42	1
70	86	5	41	76	3	12	40	1
69	86	5	40	76	3	11	37	1
68	86	5	39	75	3	10	35	1
67	85	5	38	75	3	9	32	1
66	84	4	37	74	3	8	29	1
65	84	4	36	74	3	7	26	1
64	84	4	35	73	3	6	23	1
63	83	4	34	72	3	5	20	1
62	83	4	33	71	3	4	16	1
61	83	4	32	71	3	3	13	1
60	82	4	31	70	3	2	9	1
59	82	4	30	69	3	1	5	1
58	82	4	29	68	3	0	0	1

Examination
August 2018
Algebra I

HIGH SCHOOL MATH REFERENCE SHEET

Conversions

1 inch = 2.54 centimeters

1 meter = 39.37 inches

1 mile = 5280 feet

1 mile = 1760 yards

1 mile = 1.609 kilometers

1 kilometer = 0.62 mile

1 pound = 16 ounces

1 pound = 0.454 kilogram

1 kilogram = 2.2 pounds

1 ton = 2000 pounds

1 cup = 8 fluid ounces

1 pint = 2 cups

1 quart = 2 pints

1 gallon = 4 quarts

1 gallon = 3.785 liters

1 liter = 0.264 gallon

1 liter = 1000 cubic centimeters

Formulas

Triangle	$A = \frac{1}{2}bh$
Parallelogram	$A = bh$
Circle	$A = \pi r^2$
Circle	$C = \pi d$ or $C = 2\pi r$

Formulas (continued)

General Prisms	$V = Bh$
Cylinder	$V = \pi r^2 h$
Sphere	$V = \dfrac{4}{3}\pi r^3$
Cone	$V = \dfrac{1}{3}\pi r^2 h$
Pyramid	$V = \dfrac{1}{3}Bh$
Pythagorean Theorem	$a^2 + b^2 = c^2$
Quadratic Formula	$x = \dfrac{-b \pm \sqrt{b^2 - 4ac}}{2a}$
Arithmetic Sequence	$a_n = a_1 + (n-1)d$
Geometric Sequence	$a_n = a_1 r^{n-1}$
Geometric Series	$S_n = \dfrac{a_1 - a_1 r^n}{1 - r}$ where $r \neq 1$
Radians	$1 \text{ radian} = \dfrac{180}{\pi} \text{ degrees}$
Degrees	$1 \text{ degree} = \dfrac{\pi}{180} \text{ radians}$
Exponential Growth/Decay	$A = A_0 e^{k(t - t_0)} + B_0$

PART I

Answer all 24 questions in this part. Each correct answer will receive 2 credits. No partial credit will be allowed. For each statement or question, write in the space provided the numeral preceding the word or expression that best completes the statement or answers the question. [48 credits]

1 The number of bacteria grown in a lab can be modeled by $P(t) = 300 \cdot 2^{4t}$, where t is the number of hours. Which expression is equivalent to $P(t)$?

(1) $300 \cdot 8^t$ (3) $300^t \cdot 2^4$

(2) $300 \cdot 16^t$ (4) $300^{2t} \cdot 2^{2t}$

1 _____

2 During physical education class, Andrew recorded the exercise times in minutes and heart rates in beats per minute (bpm) of four of his classmates. Which table best represents a linear model of exercise time and heart rate?

Student 1

Exercise Time (in minutes)	Heart Rate (bpm)
0	60
1	65
2	70
3	75
4	80

(1)

Student 3

Exercise Time (in minutes)	Heart Rate (bpm)
0	58
1	65
2	70
3	75
4	79

(3)

Student 2

Exercise Time (in minutes)	Heart Rate (bpm)
0	62
1	70
2	83
3	88
4	90

(2)

Student 4

Exercise Time (in minutes)	Heart Rate (bpm)
0	62
1	65
2	66
3	73
4	75

(4)

2 _____

3 David correctly factored the expression $m^2 - 12m - 64$. Which expression did he write?

(1) $(m - 8)(m - 8)$ (3) $(m - 16)(m + 4)$

(2) $(m - 8)(m + 8)$ (4) $(m + 16)(m - 4)$ 3 _____

4 The solution to $-2(1 - 4x) = 3x + 8$ is

(1) $\dfrac{6}{11}$ (3) $-\dfrac{10}{7}$

(2) 2 (4) -2 4 _____

5 The graph of $f(x)$ is shown below.

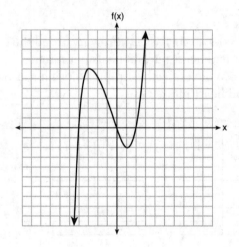

What is the value of $f(-3)$?

(1) 6   (3) -2

(2) 2   (4) -4 5 _____

6 If the function $f(x) = x^2$ has the domain $\{0, 1, 4, 9\}$, what is its range?

(1) $\{0, 1, 2, 3\}$ (3) $\{0, -1, 1, -2, 2, -3, 3\}$

(2) $\{0, 1, 16, 81\}$ (4) $\{0, -1, 1, -16, 16, -81, 81\}$ 6 _____

7 The expression $4x^2 - 25$ is equivalent to

(1) $(4x - 5)(x + 5)$ (3) $(2x + 5)(2x - 5)$

(2) $(4x + 5)(x - 5)$ (4) $(2x - 5)(2x - 5)$ 7 _____

8 Compared to the graph of $f(x) = x^2$, the graph of $g(x) = (x - 2)^2 + 3$ is the result of translating $f(x)$

(1) 2 units up and 3 units right

(2) 2 units down and 3 units up

(3) 2 units right and 3 units up

(4) 2 units left and 3 units right   8 _____

9 Lizzy has 30 coins that total $4.80. All of her coins are dimes, D, and quarters, Q. Which system of equations models this situation?

(1) $D + Q = 4.80$ (3) $D + Q = 30$

  $.10D + .25Q = 30$ $.25D + .10Q = 4.80$

(2) $D + Q = 30$ (4) $D + Q = 4.80$

  $.10D + .25Q = 4.80$ $.25D + .10Q = 30$ 9 _____

10 Gretchen has $50 that she can spend at the fair. Ride tickets cost $1.25 each and game tickets cost $2 each. She wants to go on a minimum of 10 rides and play at least 12 games.

Which system of inequalities represents this situation when r is the number of ride tickets purchased and g is the number of game tickets purchased?

(1) $1.25r + 2g < 50$ (3) $1.25r + 2g \leq 50$

  $r \leq 10$ $r \geq 10$

  $g > 12$ $g > 12$

(2) $1.25r + 2g \leq 50$ (4) $1.25r + 2g < 50$

  $r \geq 10$ $r \leq 10$

  $g \geq 12$ $g \geq 12$ 10 _____

11 Three functions are shown below.

$g(x) = 3^x + 2$

x	h(x)
-5	30
-4	14
-3	6
-2	2
-1	0
0	-1
1	-1 5
2	-1.75

Which statement is true?

(1) The y-intercept for $h(x)$ is greater than the y-intercept for $f(x)$.
(2) The y-intercept for $f(x)$ is greater than the y-intercept for $g(x)$.
(3) The y-intercept for $h(x)$ is greater than the y-intercept for both $g(x)$ and $f(x)$.
(4) The y-intercept for $g(x)$ is greater than the y-intercept for both $f(x)$ and $h(x)$.

11 _____

12 Olivia entered a baking contest. As part of the contest, she needs to demonstrate how to measure a gallon of milk if she only has a teaspoon measure. She converts the measurement using the ratios below:

$$\frac{4\ quarts}{1\ gallon} \cdot \frac{2\ pints}{1\ quart} \cdot \frac{2\ cups}{1\ pint} \cdot \frac{\frac{1}{4}cup}{4\ tablespoons} \cdot \frac{3\ teaspoons}{1\ tablespoon}$$

Which ratio is *incorrectly* written in Olivia's conversion?

(1) $\dfrac{4\ quarts}{1\ gallon}$ (3) $\dfrac{\frac{1}{4}\ cup}{4\ tablespoons}$

(2) $\dfrac{2\ pints}{1\ quart}$ (4) $\dfrac{3\ teaspoons}{1\ tablespoon}$

12 _____

13 If $y = 3x^3 + x^2 - 5$ and $z = x^2 - 12$, which polynomial is equivalent to $2(y + z)$?

(1) $6x^3 + 4x^2 - 34$ (3) $6x^3 + 3x^2 - 22$

(2) $6x^3 + 3x^2 - 17$ (4) $6x^3 + 2x^2 - 17$ 13 _____

14 An outdoor club conducted a survey of its members. The members were asked to state their preference between skiing and
snowboarding. Each member had to pick one. Of the 60 males,
45 stated they preferred to snowboard. Twenty-two of the
60 females preferred to ski. What is the relative frequency that a
male prefers to ski?

(1) 0.125   (3) $0.\overline{333}$

(2) 0.25   (4) $0.\overline{405}$ 14 _____

15 When the function $g(x) = \begin{cases} 5x, x \le 3 \\ x^2 + 4, x > 3 \end{cases}$ is graphed correctly,

how should the points be drawn on the graph for an x-value of 3?

(1) open circles at (3, 15) and (3, 13)

(2) closed circles at (3, 15) and (3, 13)

(3) an open circle at (3, 15) and a closed circle at (3, 13)

(4) a closed circle at (3, 15) and an open
circle at (3, 13)   15 _____

16 If $f(x) = 2x^2 + x - 3$, which equation can be used to determine the
zeros of the function?

(1) $0 = (2x - 3)(x + 1)$ (3) $0 = 2x(x + 1) - 3$

(2) $0 = (2x + 3)(x - 1)$ (4) $0 = 2x(x - 1) - 3(x + 1)$ 16 _____

17 Each day, a local dog shelter spends an average of $2.40 on food per dog. The manager estimates the shelter's daily expenses, assuming there is at least one dog in the shelter, using the function $E(x) = 30 + 2.40x$.

Which statements regarding the function $E(x)$ are correct?
  I. x represents the number of dogs at the shelter per day.
  II. x represents the number of volunteers at the shelter per day.
  III. 30 represents the shelter's total expenses per day.
  IV. 30 represents the shelter's nonfood expenses per day.

(1) I and III   (3) II and III
(2) I and IV   (4) II and IV   17 _____

18 Which point is not in the solution set of the equation $3y + 2 = x^2 - 5x + 17$?

(1) $(-2, 10)$ (3) $(2, 3)$
(2) $(-1, 7)$ (4) $(5, 5)$ 18 _____

19 The functions $f(x)$ and $g(x)$ are graphed below.

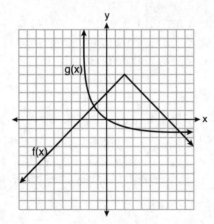

Based on the graph, the solutions to the equation $f(x) = g(x)$ are

(1) the x-intercepts
(2) the y-intercepts
(3) the x-values of the points of intersection
(4) the y-values of the points of intersection   19 _____

20 For the sequence $-27, -12, 3, 18, \ldots$, the expression that defines the nth term where $a_1 = -27$ is

(1) $15 - 27n$ (3) $-27 + 15n$

(2) $15 - 27(n - 1)$ (4) $-27 + 15(n - 1)$ 20 _____

21 The data obtained from a random sample of track athletes showed that as the foot size of the athlete decreased, the average running speed decreased. Which statement is best supported by the data?

(1) Smaller foot sizes cause track athletes to run slower.
(2) The sample of track athletes shows a causal relationship between foot size and running speed.
(3) The sample of track athletes shows a correlation between foot size and running speed.
(4) There is no correlation between foot size and running speed in track athletes.   21 _____

22 Which system of equations will yield the same solution as the system below?

$$x - y = 3$$
$$2x - 3y = -1$$

(1) $-2x - 2y = -6$ (3) $2x - 2y = 6$
     $2x - 3y = -1$ $2x - 3y = -1$
(2) $-2x + 2y = 3$ (4) $3x + 3y = 9$
     $2x - 3y = -1$ $2x - 3y = -1$ 22 _____

23 Which of the three situations given below is best modeled by an exponential function?

I. A bacteria culture doubles in size every day.
II. A plant grows by 1 inch every 4 days.
III. The population of a town declines by 5% every 3 years.

(1) I, only   (3) I and II

(2) II, only   (4) I and III   23 _____

24 The length, width, and height of a rectangular box are represented by $2x$, $3x + 1$, and $5x - 6$, respectively. When the volume is
expressed as a polynomial in standard form, what is the coefficient
of the 2nd term?

(1) -13 (3) -26

(2) 13 (4) 26 24 _____

PART II

Answer all 8 questions in this part. Each correct answer will receive 2 credits. Clearly indicate the necessary steps, including appropriate formula substitutions, diagrams, graphs, charts, etc. For all questions in this part, a correct numerical answer with no work shown will receive only 1 credit.   [16 credits]

25 Explain how to determine the zeros of
$$f(x) = (x + 3)(x - 1)(x - 8).$$

State the zeros of the function.

26 Four relations are shown below.

x	y
−4	1
0	3
4	5
6	6

I   III

$\{(1, 2), (2, 5), (3, 8), (2, -5), (1, -2)\}$ $y = x^2$

II   IV

State which relation(s) are functions.

Explain why the other relation(s) are *not* functions.

27 The table below represents the height of a bird above the ground during flight, with $P(t)$ representing height in feet and t representing time in seconds.

t	P(t)
0	6.71
3	6.26
4	6
9	3.41

Calculate the average rate of change from 3 to 9 seconds, in feet per second.

28 Is the solution to the quadratic equation written below rational or irrational? Justify your answer.

$$0 = 2x^2 + 3x - 10$$

29 The formula for converting degrees Fahrenheit (F) to degrees Kelvin (K) is:

$$K = \frac{5}{9}(F + 459.67)$$

Solve for F, in terms of K.

30 Solve the following equation by completing the square:

$$x^2 + 4x = 2$$

31 The students in Mrs. Lankford's 4th and 6th period Algebra classes took the same test. The results of the scores are shown in the following table:

	$\bar{x}$	σ_x	n	min	Q_1	med	Q_3	max
4th Period	77.75	10.79	20	58	69	76.5	87.5	96
6th Period	78.4	9.83	20	59	71.5	78	88	96

Based on these data, which class has the largest spread of test scores? Explain how you arrived at your answer.

32 Write the first five terms of the recursive sequence defined below.

$$a_1 = 0$$

$$a_n = 2(a_{n-1})^2 - 1, \text{ for } n > 1$$

PART III

Answer all 4 questions in this part. Each correct answer will receive 4 credits. Clearly indicate the necessary steps, including appropriate formula substitutions, diagrams, graphs, charts, etc. For all questions in this part, a correct numerical answer with no work shown will receive only 1 credit.   [16 credits]

33 Sarah wants to buy a snowboard that has a total cost of $580, including tax. She has already saved $135 for it. At the end of each week, she is paid $96 for babysitting and is going to save three-quarters of that for the snowboard.

Write an inequality that can be used to determine the *minimum* number of weeks Sarah needs to babysit to have enough money to purchase the snowboard.

Determine and state the *minimum* number of full weeks Sarah needs to babysit to have enough money to purchase this snowboard.

34 A car was purchased for $25,000. Research shows that the car has an average yearly depreciation rate of 18.5%.

Create a function that will determine the value, $V(t)$, of the car t years after purchase.

Determine, to the *nearest cent*, how much the car will depreciate from year 3 to year 4.

35 Graph the following system of inequalities on the set of axes below:

$$2y \geq 3x - 16$$

$$y + 2x > -2$$

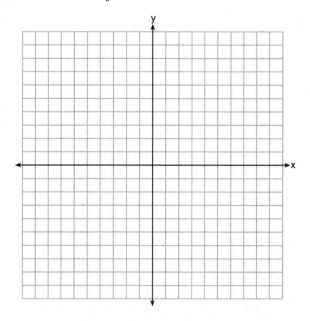

Based upon your graph, explain why $(6, 1)$ is a solution to this system and why $(-6, 7)$ is *not* a solution to this system.

36 Paul plans to have a rectangular garden adjacent to his garage. He will use 36 feet of fence to enclose three sides of the garden. The area of the garden, in square feet, can be modeled by $f(w) = w(36 - 2w)$, where w is the width in feet.

On the set of axes below, sketch the graph of $f(w)$.

Explain the meaning of the vertex in the context of the problem.

PART IV

Answer the question in this part. A correct answer will receive 6 credits. Clearly indicate the necessary steps, including appropriate formula substitutions, diagrams, graphs, charts, etc. A correct numerical answer with no work shown will receive only 1 credit. [6 credits]

37 At the present time, Mrs. Bee's age is six years more than four times her son's age. Three years ago, she was seven times as old as her son was then.

If b represents Mrs. Bee's age now and s represents her son's age now, write a system of equations that could be used to model this scenario.

Use this system of equations to determine, algebraically, the ages of both Mrs. Bee and her son now.

Determine how many years from now Mrs. Bee will be three times as old as her son will be then.

Answers
August 2018
Algebra I

Answer Key

PART I

1. (2)	**5.** (1)	**9.** (2)	**13.** (1)	**17.** (2)	**21.** (3)
2. (1)	**6.** (2)	**10.** (2)	**14.** (2)	**18.** (1)	**22.** (3)
3. (3)	**7.** (3)	**11.** (4)	**15.** (4)	**19.** (3)	**23.** (4)
4. (2)	**8.** (3)	**12.** (3)	**16.** (2)	**20.** (4)	**24.** (3)

PART II

25. $\{-3, 1, 8\}$

26. III, IV

27. -0.475

28. Irrational

29. $F = \dfrac{9}{5}K - 459.67$

30. $x = -2 \pm \sqrt{6}$

31. Period 4

32. $0, -1, 1, 1, 1$

PART III

33. $72w + 135 \geq 580$, 7 weeks

34. $V(t) = 25{,}000(0.815)^t$, 2503.71

35. $(6, 2)$ is a solution but $(-6, 7)$ is not.

36. When $w = 9$, the area is a maximum of 162.

PART IV

37. $b = 4s + 6$
$b - 3 = 7(s - 3)$
$b = 38, s = 8$, 7 years

In **Parts II–IV,** you are required to show how you arrived at your answers. For sample methods of solutions, see the *Answers Explained* section.

Answers Explained

PART I

1. There is a property of exponents that $x^{ab} = \left(x^a\right)^b$. Using this property, the expression

$$P(t) = 300 \cdot 2^{4t} = 300 \cdot \left(2^4\right)^t = 300 \cdot 16^t$$

The correct choice is (**2**).

2. Make a scatter plot for each of the four choices and find the one that looks most like a straight line.

- Choice (1):

- Choice (2):

- Choice (3):

- Choice (4):

The correct choice is (**1**).

3. To factor an expression of the form $x^2 + bx + c$ into the form $(x + p)(x + q)$, find two numbers that have a sum of b and a product of c. In this example, $b = -12$ and $c = -64$.

Since $-16 \cdot 4 = -64$ and $-16 + 4 = -12$, the two numbers are -16 and 4.

$$m^2 - 12m - 64 = (m - 16)(m + 4)$$

This question could also be answered by simplifying each of the answer choices to see which one becomes $m^2 - 12m - 64$.

- Testing choice (1):

$$(m - 8)(m - 8) = m^2 - 8m - 8m + 64 = m^2 - 16m + 64$$

- Testing choice (2):

$$(m - 8)(m + 8) = m^2 + 8m - 8m - 64 = m^2 - 64$$

- Testing choice (3):

$$(m - 16)(m + 4) = m^2 + 4m - 16m - 64 = m^2 - 12m - 64$$

- Testing choice (4):

$$(m + 16)(m - 4) = m^2 - 4m + 16m - 64 = m^2 + 12m - 64$$

The correct choice is **(3)**.

4. Use the properties of algebra to isolate the variable.

$$-2(1 - 4x) = 3x + 8$$
$$-2 + 8x = 3x + 8$$
$$-3x = -3x$$
$$-2 + 5x = 8$$
$$+2 = +2$$
$$\frac{5x}{5} = \frac{10}{5}$$
$$x = 2$$

The correct choice is **(2)**.

5. Find the point on the graph that has an x-coordinate of -3. The value of $f(-3)$ will be the y-coordinate of that point.

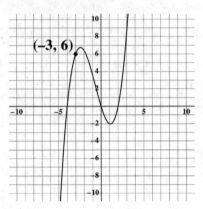

The correct choice is **(1)**.

6. Plug each of the numbers from the domain into the function. The values that the function evaluates to will be the range.

$$f(0) = 0^2 = 0$$
$$f(1) = 1^2 = 1$$
$$f(4) = 4^2 = 16$$
$$f(9) = 9^2 = 81$$

The range is the set of numbers $\{0, 1, 16, 81\}$.

The correct choice is **(2)**.

7. An expression of the form $a^2 - b^2$ factors into $(a - b)(a + b)$. The expression $4x^2 - 25$ can be written as $(2x)^2 - 5^2$. Using the factoring pattern, this becomes $(2x)^2 - 5^2 = (2x - 5)(2x + 5) = (2x + 5)(2x - 5)$.

This question could also be answered by simplifying each of the answer choices to see which one becomes $4x^2 - 25$.

- Testing choice (1):

$$(4x - 5)(x + 5) = 4x^2 + 20x - 5x - 25 = 4x^2 + 15x - 25$$

- Testing choice (2):

$$(4x + 5)(x - 5) = 4x^2 - 20x + 5x - 25 = 4x^2 - 15x - 25$$

- Testing choice (3):

$$(2x + 5)(2x - 5) = 4x^2 - 10x + 10x - 25 = 4x^2 - 25$$

- Testing choice (4):

$$(2x - 5)(2x - 5) = 4x^2 - 10x - 10x + 25 = 4x^2 - 20x + 25$$

The correct choice is **(3)**.

8. On your graphing calculator, graph both functions.

The graph of $g(x)$ is a translation of the graph of $f(x)$ two units to the right and three units up. This can also be answered without a graphing calculator.

In general:

The graph of $f(x + a)$ is a translation of the graph of $f(x)$ by a units to the left.

The graph of $f(x - a)$ is a translation of the graph of $f(x)$ by a units to the right.

The graph of $f(x) + a$ is a translation of the graph of $f(x)$ by a units up.

The graph of $f(x) - a$ is a translation of the graph of $f(x)$ by a units down.

In this example, $g(x) = f(x - 2) + 3$, so the graph is a translation of the graph $f(x)$ two units to the right and three units up.

The correct choice is **(3)**.

9. Since D is the number of dimes and Q is the number of quarters, the total number of coins is $D + Q$. Since there are 30 coins, the first equation is $D + Q = 30$.

   Since each dime is worth 10 cents, the value of D dimes is $.10D$. Since each quarter is worth 25 cents, the value of Q quarters is $.25Q$. The total value of the coins is $.10D + .25Q$. Since the coins are worth $4.80, the second equation is $.10D + .25Q = 4.80$.

   The correct equation is (**2**).

10. Since rides cost $1.25 each, the cost of r rides is $1.25r$. Since games cost $2 each, the cost of g games is $2g$.

    Since she has $50, the first equation is $1.25r + 2g \leq 50$. It must be a $\leq$, not a $<$, because, if the total cost is $50, she can afford it. She cannot spend more than $50, but she can spend exactly $50 if that's what it costs.

    Since she wants to go on a minimum of 10 rides, the second equation is $r \geq 10$.

    Since she wants to play a minimum of 12 games, the third equation is $g \geq 12$.

    The correct choice is (**2**).

11. The y-intercept of $f(x)$ is the y-coordinate of the point where the curve crosses the y-axis. Since the curve crosses the y-axis at the point $(0, 1)$, the y-intercept is 1.

    The y-intercept of the graph of $g(x)$ is the value of $g(0)$.

    For this function $g(0) = 3^0 + 2 = 1 + 2 = 3$, so the y-intercept for $g(x)$ is 3.

    The y-intercept of the graph of $h(x)$ is the value of $h(0)$.

    According to the chart, $h(0) = -1$, so the y-intercept for $h(x)$ is -1.

    Since $3 > 1 > -1$, the y-intercept of $g(x)$ is greater than the y-intercept for both $f(x)$ and $h(x)$.

    The correct choice is (**4**).

12. When doing unit conversions, make the denominator of each fraction have the same kind of units as the numerator in the previous fraction.

The units of the second denominator are quarts, which match the units of the first numerator.

The units of the third denominator are pints, which match the units of the second numerator.

The units of the fourth denominator are tablespoons, which do not match the units of the third numerator.

This fourth fraction would need to be "flipped" to do the unit conversion properly.

Note: The conversions of tablespoons to cups (16 tablespoons = 1 cup) and teaspoons to tablespoons (3 teaspoons = 1 tablespoon) are not on the reference sheet. Still, this question can be answered by using the above strategy while assuming that the conversion rates are accurate.

The correct choice is (**3**).

13. Replace y and z with the equivalent expressions in terms of x.

$$2(y + z) = 2(3x^3 + x^2 - 5 + x^2 - 12) = 2(3x^3 + 2x^2 - 17)$$
$$= 6x^3 + 4x^2 - 34$$

The correct choice is (**1**).

14. The data can first be organized into a two-way frequency table:

	Skiing	**Snowboarding**
Male	15	45
Female	22	38

Out of the 60 males, 15 of them prefer to ski, so the relative frequency that a male prefers to ski is $\dfrac{15}{60} = 0.25$.

The correct choice is (**2**).

15. The graph of this piecewise function looks like this:

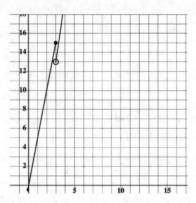

Since the first "piece" has the $\leq$ sign, that piece will include the endpoint with the closed circle at $(3, 15)$.

Since the second "piece" has the $>$ sign, that piece will not contain the endpoint of $(3, 13)$, so there will be an open circle there.

The correct choice is **(4)**.

16. The zeros of a function, $f(x)$, are the values for x that make $f(x) = 0$.

Solving the equation:

$$f(x) = 0$$
$$2x^2 + x - 3 = 0$$

The next step is to factor the quadratic equation:

$$(2x + 3)(x - 1) = 0$$

This question does not ask you to solve for x, but just to find the equation that can be used to solve for x.

The correct choice is **(2)**.

17. One way to analyze the components of this equation is to calculate the values of $E(1)$ and $E(2)$.

Plugging into the function you get:

$$E(1) = 30 + 2.40 \cdot 1 = 30 + 2.40 = 32.40$$
$$E(2) = 30 + 2.40 \cdot 2 = 30 + 4.80 = 34.80$$

Each time x increases by 1, $E(x)$ increases by 2.40, so x must represent the number of dogs at the shelter per day; therefore, Roman numeral I is one of the correct statements.

This eliminates choices (3) and (4).

Roman numeral III cannot be correct because the total expenses change depending on the value of x, so 30 must represent the shelter's nonfood expenses per day, which is the statement in Roman numeral IV.

The correct choice is **(2)**.

18. Any ordered pair in the solution set of an equation will make the left-hand side equal the right-hand side.

- Testing choice (1):

$$3 \cdot 10 + 2 \overset{?}{=} (-2)^2 - 5(-2) + 17$$
$$30 + 2 \overset{?}{=} 4 + 10 + 17$$
$$32 \neq 31$$

- Testing choice (2):

$$3 \cdot 7 + 2 \overset{?}{=} (-1)^2 - 5(-1) + 17$$
$$21 + 2 \overset{?}{=} 1 + 5 + 17$$
$$23 \overset{\checkmark}{=} 23$$

- Testing choice (3):

$$3 \cdot 3 + 2 \overset{?}{=} 2^2 - 5 \cdot 2 + 17$$
$$9 + 2 \overset{?}{=} 4 - 10 + 17$$
$$11 \overset{\checkmark}{=} 11$$

- Testing choice (4):

$$3 \cdot 5 + 2 \overset{?}{=} 5^2 - 5 \cdot 5 + 17$$
$$15 + 2 \overset{?}{=} 25 - 25 + 17$$
$$17 \overset{\checkmark}{=} 17$$

Of the four choices, only choice (1) is not in the solution set.

The correct choice is **(1)**.

19. A solution to this equation is an x-value that is put into the functions that will evaluate to the same output value for both functions. The graphs of these two functions intersect at approximately $(-1.5, 1.5)$ and $(8.5, -1.5)$.

    This means that $f(-1.5) = g(-1.5)$ and $f(8.5) = g(8.5)$. These solutions are the x-coordinates of the points of intersection.

    The correct choice is **(3)**.

20. An arithmetic sequence is one where each term is the same amount more (or less) than the previous term. For this sequence, the first term of this sequence is -27, and each term is 15 more than the previous term.

    There is a formula for the nth term of an arithmetic sequence in the reference sheet provided on the test.

    $a_n = a_1 + (n - 1)d$, where a_1 is the first term, n is the term number, and d is the "common difference" (what you need to add to a term to get the next term).

    For this sequence, $a_1 = -27$ and $d = +15$, so the formula is

    $$a_n = -27 + (n - 1)15 = -27 + 15(n - 1)$$

    The correct choice is **(4)**.

21. Since the running speed decreases when the foot size decreases, you can say, for sure, that there is a correlation between the two things. It may be true, also, that the smaller foot size actually causes the slower speed, but you cannot be sure about this. The smaller foot size and the slower speed could, for example, both be caused by some third thing, like the runner being a child. All you can say with certainty is that there is a correlation between the two things.

    The correct choice is **(3)**.

22. When you have a system of equations, the solution will not change if you multiply both sides of either equation by some number. Test each of the choices to see if that system can be obtained from the original system by multiplying both sides of either one or both equations by some number.

    - Testing choice (1): If you multiply both sides of the top equation by -2, you would get $-2x + 2y = -6$, *not* $-2x - 2y = -6$.
    - Testing choice (2): If you multiply both sides of the top equation by -2, you would get $-2x + 2y = -6$, *not* $-2x + 2y = 3$.
    - Testing choice (3): This is the system you would get if you multiply both sides of the top equation by 2.

- Testing choice (4): If you multiply both sides of the top equation by 3, you would get $3x - 3y = 9$, *not* $3x + 3y = 9$.

The correct choice is **(3)**.

23. An exponential function has the form $f(x) = a \cdot b^x$. The functions for each of the Roman numerals are:

- Scenario I: $f(x) = a \cdot 2^x$
- Scenario II: $f(x) = \dfrac{1}{4}x$
- Scenario III: $f(x) = a(1 - 0.05)^{x/3}$

Scenarios I and III have the form of an exponential function.

Even without creating the equations, you can make a rough sketch of the graph of each scenario to see which has the shape of an exponential curve (which looks a little like a playground slide).

Graph of scenario 1:

Graph of scenario 2:

Graph of scenario 3:

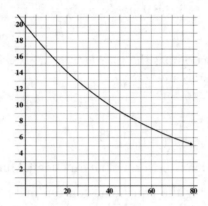

The correct choice is **(4)**.

24. The volume of a rectangular box can be calculated with the formula $V = lwh$, where l is the length, w is the width, and h is the height.

    For this question, this becomes $V = (2x)(3x + 1)(5x - 6)$.

    Multiply the factors of the right-hand side to turn it into a polynomial.

    $$(2x)(3x + 1)(5x - 6)$$
    $$(6x^2 + 2x)(5x - 6)$$
    $$30x^3 - 36x^2 + 10x^2 - 12x$$
    $$30x^3 - 26x^2 - 12x$$

    For a polynomial to be in standard form, the first term must be the one with the largest exponent, and the size of the exponent must decrease as you go from left to right. Since the exponent of 3 is before the exponent of 2, which is before the exponent of 1, this is a polynomial in standard form.

    The coefficient of the second term is -26.

    The correct choice is **(3)**.

PART II

25. The zeros of a function, $f(x)$, are the values of x that make the function evaluate to 0.

    To determine the zeros, solve the equation $f(x) = 0$.

    $$0 = (x + 3)(x - 1)(x - 8)$$

    The only way the product of three different things can be zero is if one of those things is zero.

    $$\text{If } (x + 3) = 0, \text{ then } x = -3.$$
    $$\text{If } (x - 1) = 0, \text{ then } x = 1.$$
    $$\text{If } (x - 8) = 0, \text{ then } x = 8.$$

    So, the zeros of the function are -3, 1, and 8.

26. For Roman numeral I, the graph of a function will not have more than one point with the same x-coordinate. Since there is a point at $(6, 4)$ and also a point at $(6, 6)$, this cannot be a function.

    A graph where this happens is said to fail the "vertical line test," since it is possible to draw a vertical line that would pass through more than one point on the graph.

    For Roman numeral II, check the list to see if there are any ordered pairs that have the same x-coordinate. Since there are repeated x-values in the ordered pairs $(1, 2)$ and $(1, -2)$ and also repeated x-values in the ordered pairs $(2, 5)$ and $(2, -5)$, this is not a function.

    For Roman numeral III, check to see if there are two rows in the chart that have the same x-value. Since all the x-values are different, this is a function.

For Roman numeral IV, check to see if the graph of the equation passes the vertical line test. Since there is no vertical line that would pass through more than one point on the graph of this equation, this is a function.

Relations III and IV are functions, while relations I and II are not functions.

27. The average rate of change of a function $P(t)$ on the interval from a to b can be calculated by the formula:

$$\text{Average Rate of Change} = \frac{P(b) - P(a)}{b - a}$$

For this example,

$$\text{Average Rate of Change} = \frac{P(9) - P(3)}{9 - 3} = \frac{3.41 - 6.26}{9 - 3} = \frac{-2.85}{6}$$
$$= -0.475$$

28. The solutions to a quadratic equation $ax^2 + bx + c = 0$ are rational if the "discriminant" $D = b^2 - 4ac$ is a perfect square.

For this example, $D = 3^2 - 4 \cdot 2 \cdot (-10) = 9 + 80 = 89$. Since 89 is not a perfect square, the solutions to this quadratic equation are irrational.

Another way to answer this question is to solve the equation using the quadratic formula to show that both answers are irrational.

$$x = \frac{-b \pm \sqrt{b^2 - 4ac}}{2a} = \frac{-3 \pm \sqrt{3^2 - 4 \cdot 2 \cdot (-10)}}{2 \cdot 2} = \frac{-3 \pm \sqrt{9 + 80}}{4}$$
$$= \frac{-3 \pm \sqrt{89}}{4}$$

Since 89 is not a perfect square, $\sqrt{89}$ is irrational, and so are both $\dfrac{-3 + \sqrt{89}}{4}$ and $\dfrac{-3 - \sqrt{89}}{4}$.

29. The simplest way to isolate the F is to first multiply both sides of the equation by the reciprocal of $\frac{5}{9}$, which is $\frac{9}{5}$.

$$K = \frac{5}{9}(F + 459.67)$$

$$\frac{9}{5}K = \frac{9}{5} \cdot \frac{5}{9}(F + 459.67)$$

$$\frac{9}{5}K = 1(F + 459.67)$$

$$\frac{9}{5}K = F + 459.67$$

$$-459.67 = -459.67$$

$$\frac{9}{5}K - 459.67 = F$$

30. To solve a quadratic equation in the form $x^2 + bx = c$ by the "completing the square" process, the first step is to add $\left(\frac{b}{2}\right)^2$ to both sides of the equation.

$$x^2 + 4x = 2$$

$$+\left(\frac{4}{2}\right)^2 = +\left(\frac{4}{2}\right)^2$$

$$x^2 + 4x + 4 = 2 + 4$$

$$x^2 + 4x + 4 = 6$$

The left side of this equation is now a perfect square and can be factored:

$$x^2 + 4x + 4 = 6$$

$$(x + 2)(x + 2) = 6$$

$$(x + 2)^2 = 6$$

Finish by taking the square root of both sides of the equation and then isolating the x.

$$\sqrt{(x + 2)^2} = \pm\sqrt{6}$$

$$x + 2 = \pm\sqrt{6}$$

$$-2 = -2$$

$$x = -2 \pm \sqrt{6}$$

31. The "spread" is the difference between the maximum score and the minimum score.

    The spread for 4th period is $96 - 58 = 38$.

    The spread for 6th period is $96 - 59 = 37$.

    Since $38 > 37$, the class with the largest spread is period 4.

32. The first term of the sequence is given as $a_1 = 0$. To calculate the next four terms, follow the second rule for $n = 2, 3, 4,$ and 5.

    For $n = 2$,
    $$a_2 = 2(a_{2-1})^2 - 1 = 2(a_1)^2 - 1 = 2 \cdot 0^2 - 1 = 2 \cdot 0 - 1 = 0 - 1 = -1$$

    For $n = 3$,
    $$a_3 = 2(a_{3-1})^2 - 1 = 2(a_2)^2 - 1 = 2(-1)^2 - 1 = 2 \cdot 1 - 1 = 2 - 1 = 1$$

    For $n = 4$,
    $$a_4 = 2(a_{4-1})^2 - 1 = 2(a_3)^2 - 1 = 2 \cdot 1^2 - 1 = 2 \cdot 1 - 1 = 2 - 1 = 1$$

    For $n = 5$,
    $$a_5 = 2(a_{5-1})^2 - 1 = 2(a_4)^2 - 1 = 2 \cdot 1^2 - 1 = 2 \cdot 1 - 1 = 2 - 1 = 1$$

    The first five terms are $0, -1, 1, 1, 1$.

PART III

33. Each week she saves $\frac{3}{4} \times \$96 = \$72$. In w weeks, then, she saves $72w$ dollars. Since she already saved $135, after w weeks she will have saved a total of $72w + 135$ dollars.

    To find the minimum number of weeks she must work to save at least $580, the inequality is:

    $$72w + 135 \geq 580$$

    Solve as you would an equality.

    $$72w + 135 \geq 580$$
    $$-135 = -135$$
    $$\frac{72w}{72} \geq \frac{445}{72}$$
    $$w \geq 6.18$$

    Since she must work a full week, because 6 weeks will not be enough, the minimum number of full weeks she must work is 7.

34. This is an example of exponential decay. The formula for exponential decay is $A = P(1 - r)^t$, where P is the starting value, r is the decay rate, and t is the amount of time.

    For this example, the formula is

    $$V(t) = 25,000(1 - 0.185)^t = 25,000(0.815)^t$$

    The value of the car at $t = 3$ is

    $$V(3) = 25,000(0.815)^3 \approx 13,533.58$$

    The value of the car at $t = 4$ is

    $$V(4) = 25,000(0.815)^4 \approx 11,029.87$$

    The difference between these two values is

    $$13,533.58 - 11,029.87 = 2503.71$$

35. Each of the inequalities will be graphed as a boundary line with one side of the line shaded.

To graph the boundary line for the first inequality, change the $\geq$ to an $=$ and convert to slope-intercept form $(y = mx + b)$.

$$2y = 3x - 16$$

$$\frac{2y}{2} = \frac{3x - 16}{2}$$

$$y = \frac{3}{2}x - 8$$

Since the inequality symbol was a $\geq$, graph this line as a solid line. If it were a $>$, you would graph it as a dotted line.

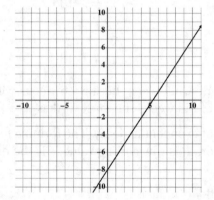

To determine which side to shade, check to see if the ordered pair (0, 0) makes the inequality true or false.

$$2 \cdot 0 \geq 3 \cdot 0 - 16$$

$$0 \geq 0 - 16$$

$$0 \geq -16$$

Since it is true that $0 \geq -16$, shade the side of the line that contains $(0, 0)$.

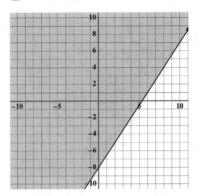

For the second inequality, find the equation of the boundary line the same way.

$$y + 2x = -5$$
$$-2x = -2x$$
$$y = -2x - 5$$

Since the inequality symbol for the second inequality was a $>$, make this line a dotted line.

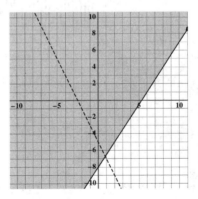

To determine which side to shade, check to see if the ordered pair $(0, 0)$ makes the second inequality true or false.

$$0 + 2 \cdot 0 > -5$$
$$0 + 0 > -5$$
$$0 > -5$$

Since it is true that $0 > -5$, shade the side of the line that contains $(0, 0)$.

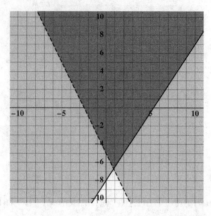

The inequality graphs can also be made on the graphing calculator.

All the points in the double-shaded area are solutions to the system of inequalities. Since (6, 1) is on the solid boundary line, it is considered in the double-shaded area, but (−6, 7) is not in the double-shaded area because it is on the dotted line.

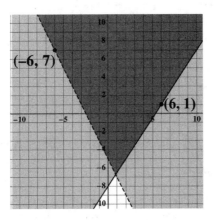

36. The simplest way to make the graph is to create a chart.

w	$f(w)$
0	$0(36 - 0) = 0$
1	$1(36 - 2 \cdot 1) = 34$
2	$2(36 - 2 \cdot 2) = 64$
3	$3(36 - 2 \cdot 3) = 90$
4	$4(36 - 2 \cdot 4) = 112$
5	$5(36 - 2 \cdot 5) = 130$
6	$6(36 - 2 \cdot 6) = 144$
7	$7(36 - 2 \cdot 7) = 154$
8	$8(36 - 2 \cdot 8) = 160$
9	$9(36 - 2 \cdot 9) = 162$

w	$f(w)$
10	$10(36 - 2 \cdot 10) = 160$
11	$11(36 - 2 \cdot 11) = 154$
12	$12(36 - 2 \cdot 12) = 144$
13	$13(36 - 2 \cdot 13) = 130$
14	$14(36 - 2 \cdot 14) = 112$
15	$15(36 - 2 \cdot 15) = 90$
16	$16(36 - 2 \cdot 16) = 64$
17	$17(36 - 2 \cdot 17) = 34$
18	$18(36 - 2 \cdot 18) = 0$

The graph looks like this:

The graph can also be created on a graphing calculator.

The shape of the graph is a parabola. The vertex of this parabola is the highest point, which is (9, 162). The meaning of this vertex is that if he makes the width 9 feet, the area of the garden will be 162 square feet, which is the greatest possible area with the fencing he has.

37. Mrs. Bee's age now is represented by b. Her son's age now is represented by s.

The first equation is $b = 4s + 6$.

Three years ago, Mrs. Bee's age was $b - 3$. Her son's age, three years ago, was $s - 3$.

The second equation is $b - 3 = 7(s - 3)$.

The system of equations is:

$$b = 4s + 6$$
$$b - 3 = 7(s - 3)$$

Using substitution, replace each b in the second equation with the expression $4s + 6$.

$$4s + 6 - 3 = 7(s - 3)$$
$$4s + 3 = 7(s - 3)$$
$$4s + 3 = 7s - 21$$
$$-4s = -4s$$
$$3 = 3s - 21$$
$$+21 = +21$$
$$\frac{24}{3} = \frac{3s}{3}$$
$$8 = s$$
$$b = 4 \cdot 8 + 6 = 38$$

The son is now 8 years old and Mrs. Bee is now 38 years old.

In x years, Mrs. Bee will be three times as old as her son will be then. In x years, Mrs. Bee will be $38 + x$ years old. In x years, her son will be $8 + x$ years old.

This becomes the equation $38 + x = 3(8 + x)$.

Solve for x:

$$38 + x = 3(8 + x)$$
$$38 + x = 24 + 3x$$
$$-x = -x$$
$$38 = 24 + 2x$$
$$-24 = -24$$
$$\frac{14}{2} = \frac{2x}{2}$$
$$7 = x$$

In 7 years, Mrs. Bee will be $38 + 7 = 45$ years old. Her son will be $8 + 7 = 15$ years old. 45 is three times 15.

Topic	Question Numbers	Number of Points
1. Polynomials	4, 13, 25, 15, 29	$2 + 2 + 2 + 2 + 2 = 10$
2. Properties of Algebra		0
3. Functions	2, 5, 6, 11, 19, 26, 27	$2 + 2 + 2 + 2 + 2 + 2 + 2 = 14$
4. Creating and Interpreting Equations	17, 23, 24	$2 + 2 + 4 = 8$
5. Inequalities	10, 33	$2 + 4 = 6$
6. Sequences and Series	20, 32	$2 + 2 = 4$
7. Systems of Equations	9, 22, 37	$2 + 2 + 6 = 10$
8. Quadratic Equations and Factoring	3, 7, 16, 30	$2 + 2 + 2 + 2 = 8$
9. Regression		0
10. Exponential Equations	1	2
11. Graphing	8, 15, 18, 35, 36	$2 + 2 + 2 + 4 + 4 = 14$
12. Statistics	14, 21, 31	$2 + 2 + 2 = 6$
13. Number Properties	28	2
14. Unit Conversions	12	2

HOW TO CONVERT YOUR RAW SCORE TO YOUR ALGEBRA I REGENTS EXAMINATION SCORE

The accompanying conversion chart must be used to determine your final score on the August 2018 Regents Examination in Algebra I. To find your final exam score, locate in the column labeled "Raw Score" the total number of points you scored out of a possible 86 points. Since partial credit is allowed in Parts II, III, and IV of the test, you may need to approximate the credit you would receive for a solution that is not completely correct. Then locate in the adjacent column to the right the scale score that corresponds to your raw score. The scale score is your final Algebra I Regents Examination score.

Regents Examination in Algebra I—August 2018
Chart for Converting Total Test Raw Scores to Final Examination Scores (Scaled Scores)

Raw Score	Scale Score	Performance Level	Raw Score	Scale Score	Performance Level	Raw Score	Scale Score	Performance Level
86	100	5	57	81	4	28	66	3
85	98	5	56	81	4	27	65	3
84	97	5	55	81	4	26	64	2
83	96	5	54	81	4	25	63	2
82	94	5	53	80	4	24	61	2
81	93	5	52	80	4	23	60	2
80	92	5	51	80	4	22	58	2
79	92	5	50	80	4	21	57	2
78	91	5	49	79	3	20	55	2
77	90	5	48	79	3	19	53	1
76	89	5	47	79	3	18	51	1
75	88	5	46	78	3	17	49	1
74	88	5	45	78	3	16	47	1
73	87	5	44	78	3	15	45	1
72	87	5	43	77	3	14	43	1
71	86	5	42	77	3	13	40	1
70	86	5	41	76	3	12	38	1
69	86	5	40	76	3	11	36	1
68	85	5	39	75	3	10	33	1
67	84	4	38	75	3	9	30	1
66	84	4	37	74	3	8	27	1
65	84	4	36	74	3	7	24	1
64	83	4	35	73	3	6	21	1
63	83	4	34	72	3	5	18	1
62	83	4	33	71	3	4	15	1
61	82	4	32	71	3	3	11	1
60	82	4	31	70	3	2	8	1
59	82	4	30	69	3	1	4	1
58	82	4	29	68	3	0	0	1

Examination
June 2019
Algebra I

HIGH SCHOOL MATH REFERENCE SHEET

Conversions

1 inch = 2.54 centimeters

1 meter = 39.37 inches

1 mile = 5280 feet

1 mile = 1760 yards

1 mile = 1.609 kilometers

1 kilometer = 0.62 mile

1 pound = 16 ounces

1 pound = 0.454 kilogram

1 kilogram = 2.2 pounds

1 ton = 2000 pounds

1 cup = 8 fluid ounces

1 pint = 2 cups

1 quart = 2 pints

1 gallon = 4 quarts

1 gallon = 3.785 liters

1 liter = 0.264 gallon

1 liter = 1000 cubic centimeters

Formulas

Triangle	$A = \frac{1}{2}bh$
Parallelogram	$A = bh$
Circle	$A = \pi r^2$
Circle	$C = \pi d$ or $C = 2\pi r$

Formulas (continued)

General Prisms	$V = Bh$
Cylinder	$V = \pi r^2 h$
Sphere	$V = \frac{4}{3}\pi r^3$
Cone	$V = \frac{1}{3}\pi r^2 h$
Pyramid	$V = \frac{1}{3}Bh$
Pythagorean Theorem	$a^2 + b^2 = c^2$
Quadratic Formula	$x = \dfrac{-b \pm \sqrt{b^2 - 4ac}}{2a}$
Arithmetic Sequence	$a_n = a_1 + (n-1)d$
Geometric Sequence	$a_n = a_1 r^{n-1}$
Geometric Series	$S_n = \dfrac{a_1 - a_1 r^n}{1 - r}$ where $r \neq 1$
Radians	$1 \text{ radian} = \dfrac{180}{\pi} \text{ degrees}$
Degrees	$1 \text{ degree} = \dfrac{\pi}{180} \text{ radians}$
Exponential Growth/Decay	$A = A_0 e^{k(t - t_0)} + B_0$

PART I

Answer all 24 questions in this part. Each correct answer will receive 2 credits. No partial credit will be allowed. For each statement or question, write in the space provided the numeral preceding the word or expression that best completes the statement or answers the question.   [48 credits]

1 The expression $w^4 - 36$ is equivalent to
(1) $(w^2 - 18)(w^2 - 18)$ (3) $(w^2 - 6)(w^2 - 6)$
(2) $(w^2 + 18)(w^2 - 18)$ (4) $(w^2 + 6)(w^2 - 6)$ 1 ____

2 If $f(x) = 4x + 5$, what is the value of $f(-3)$?
(1) -2 (3) 17
(2) -7 (4) 4   2 ____

3 Which relation is *not* a function?

x	y
−10	−2
−6	2
−2	6
1	9
5	13

(1)

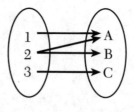

(3)

$3x + 2y = 4$

(2)

(4)   3 ____

4 Given: $f(x) = (x - 2)^2 + 4$
$g(x) = (x - 5)^2 + 4$

When compared to the graph of $f(x)$, the graph of $g(x)$ is

(1) shifted 3 units to the left
(2) shifted 3 units to the right
(3) shifted 5 units to the left
(4) shifted 5 units to the right   4 _____

5 Students were asked to write $6x^5 + 8x - 3x^3 + 7x^7$ in standard form.
Shown below are four student responses.

$$\text{Anne:} \quad 7x^7 + 6x^5 - 3x^3 + 8x$$
$$\text{Bob:} \quad -3x^3 + 6x^5 + 7x^7 + 8x$$
$$\text{Carrie:} \quad 8x + 7x^7 + 6x^5 - 3x^3$$
$$\text{Dylan:} \quad 8x - 3x^3 + 6x^5 + 7x^7$$

Which student is correct?

(1) Anne   (3) Carrie
(2) Bob   (4) Dylan   5 _____

6 The function f is shown in the table below.

x	f(x)
0	1
1	3
2	9
3	27

Which type of function best models the given data?

(1) exponential growth function
(2) exponential decay function
(3) linear function with positive rate of change
(4) linear function with negative rate of change   6 _____

7 Which expression results in a rational number?

(1) $\sqrt{2} \cdot \sqrt{18}$ (3) $\sqrt{2} + \sqrt{2}$

(2) $5 \cdot \sqrt{5}$ (4) $3\sqrt{2} + 2\sqrt{3}$ 7 _____

8 A polynomial function is graphed below.

f(x)

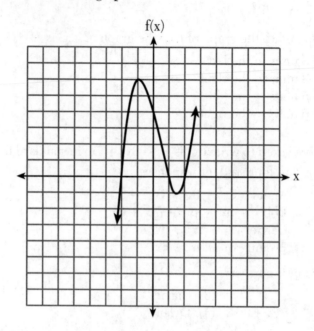

Which function could represent this graph?

(1) $f(x) = (x + 1)(x^2 + 2)$
(2) $f(x) = (x - 1)(x^2 - 2)$
(3) $f(x) = (x - 1)(x^2 - 4)$
(4) $f(x) = (x + 1)(x^2 + 4)$ 8 _____

9 When solving $p^2 + 5 = 8p - 7$, Kate wrote $p^2 + 12 = 8p$. The property she used is

(1) the associative property
(2) the commutative property
(3) the distributive property
(4) the addition property of equality   9 _____

10 David wanted to go on an amusement park ride. A sign posted at the entrance read "You must be greater than 42 inches tall and no more than 57 inches tall for this ride." Which inequality would model the height, x, required for this amusement park ride?

(1) $42 < x \leq 57$ (3) $42 < x$ or $x \leq 57$

(2) $42 > x \geq 57$ (4) $42 > x$ or $x \geq 57$ 10 _____

11 Which situation can be modeled by a linear function?

(1) The population of bacteria triples every day.

(2) The value of a cell phone depreciates at a rate of 3.5% each year.

(3) An amusement park allows 50 people to enter every 30 minutes.

(4) A baseball tournament eliminates half of the teams after each round.   11 _____

12 Jenna took a survey of her senior class to see whether they preferred pizza or burgers. The results are summarized in the table below.

	Pizza	**Burgers**
Male	23	42
Female	31	26

Of the people who preferred burgers, approximately what percentage were female?

(1) 21.3   (3) 45.6

(2) 38.2   (4) 61.9   12 _____

13 When $3a + 7b > 2a - 8b$ is solved for a, the result is

(1) $a > -b$ (3) $a < -15b$

(2) $a < -b$ (4) $a > -15b$ 13 _____

14 Three functions are shown below.

A: $g(x) = -\dfrac{3}{2}x + 4$

B: $f(x) = (x + 2)(x + 6)$

C:

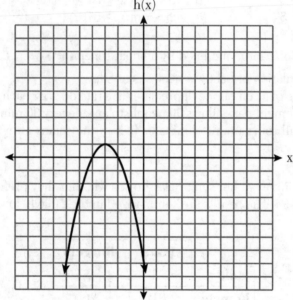

Which statement is true?

(1) B and C have the same zeros.
(2) A and B have the same y-intercept.
(3) B has a minimum and C has a maximum.
(4) C has a maximum and A has a minimum.

14 _____

15 Nicci's sister is 7 years less than twice Nicci's age, a. The sum of Nicci's age and her sister's age is 41. Which equation represents this relationship?

(1) $a + (7 - 2a) = 41$ (3) $2a - 7 = 41$
(2) $a + (2a - 7) = 41$ (4) $a = 2a - 7$

15 _____

16 The population of a small town over four years is recorded in the chart below, where 2013 is represented by $x = 0$. [Population is rounded to the nearest person]

Year	2013	2014	2015	2016
Population	3810	3943	4081	4224

The population, $P(x)$, for these years can be modeled by the function $P(x) = ab^x$, where b is rounded to the nearest thousandth. Which statements about this function are true?

    I. $a = 3810$
    II. $a = 4224$
    III. $b = 0.035$
    IV. $b = 1.035$

(1) I and III   (3) II and III
(2) I and IV   (4) II and IV   16 _____

17 When written in factored form, $4w^2 - 11w - 3$ is equivalent to

(1) $(2w + 1)(2w - 3)$ (3) $(4w + 1)(w - 3)$
(2) $(2w - 1)(2w + 3)$ (4) $(4w - 1)(w + 3)$ 17 _____

18 Which ordered pair does *not* represent a point on the graph of $y = 3x^2 - x + 7$?

(1) $(-1.5, 15.25)$ (3) $(1.25, 10.25)$
(2) $(0.5, 7.25)$ (4) $(2.5, 23.25)$ 18 _____

19 Given the following three sequences:

    I.   2, 4, 6, 8, 10...
    II.   2, 4, 8, 16, 32...
    III. $a, a + 2, a + 4, a + 6, a + 8...$

Which ones are arithmetic sequences?

(1) I and II, only   (3) II and III, only
(2) I and III, only   (4) I, II, and III   19 _____

20 A grocery store sells packages of beef. The function $C(w)$ represents the cost, in dollars, of a package of beef weighing w pounds. The most appropriate domain for this function would be

(1) integers
(2) rational numbers
(3) positive integers
(4) positive rational numbers

20 _____

21 The roots of $x^2 - 5x - 4 = 0$ are

(1) 1 and 4   (3) -1 and -4

(2) $\dfrac{5 \pm \sqrt{41}}{2}$ (4) $\dfrac{-5 \pm \sqrt{41}}{2}$

21 _____

22 The following table shows the heights, in inches, of the players on the opening-night roster of the 2015–2016 New York Knicks.

84	80	87	75	77	79	80	74	76	80	80	82	82

The population standard deviation of these data is approximately

(1) 3.5   (3) 79.7
(2) 13   (4) 80

22 _____

23 A population of bacteria can be modeled by the function $f(t) = 1000(0.98)^t$, where t represents the time since the population started decaying, and $f(t)$ represents the population of the remaining bacteria at time t. What is the rate of decay for this population?

(1) 98%   (3) 0.98%
(2) 2%   (4) 0.02%

23 _____

24 Bamboo plants can grow 91 centimeters per day. What is the approximate growth of the plant, in inches per hour?

(1) 1.49   (3) 9.63
(2) 3.79   (4) 35.83

24 _____

PART II

Answer all 8 questions in this part. Each correct answer will receive 2 credits. Clearly indicate the necessary steps, including appropriate formula substitutions, diagrams, graphs, charts, etc. For all questions in this part, a correct numerical answer with no work shown will receive only 1 credit. [16 credits]

25 Solve algebraically for x: $-\dfrac{2}{3}(x + 12) + \dfrac{2}{3}x = -\dfrac{5}{4}x + 2$

26 If $C = G - 3F$, find the trinomial that represents C when $F = 2x^2 + 6x - 5$ and $G = 3x^2 + 4$.

27 Graph the following piecewise function on the set of axes below.

$$f(x) = \begin{cases} |x|, & -5 \le x < 2 \\ -2x + 10, & 2 \le x \le 6 \end{cases}$$

f(x)

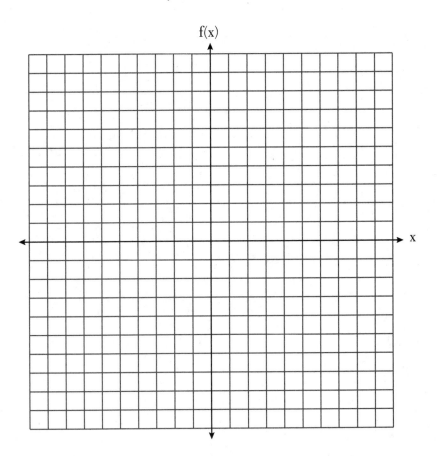

28   Solve $5x^2 = 180$ algebraically.

29 A blizzard occurred on the East Coast during January, 2016. Snowfall totals from the storm were recorded for Washington, D.C. and are shown in the table below.

Washington, D.C.	
Time	**Snow** (inches)
1 a.m.	1
3 a.m.	5
6 a.m.	11
12 noon	33
3 p.m.	36

Which interval, 1 a.m. to 12 noon or 6 a.m. to 3 p.m., has the greatest rate of snowfall, in inches per hour? Justify your answer.

30 The formula for the volume of a cone is $V = \frac{1}{3}\pi r^2 h$. Solve the equation for h in terms of V, r, and π.

31 Given the recursive formula:

$$a_1 = 3$$

$$a_n = 2(a_{n-1} + 1)$$

State the values of a_2, a_3, and a_4 for the given recursive formula.

32 Determine and state the vertex of $f(x) = x^2 - 2x - 8$ using the method of completing the square.

PART III

Answer all 4 questions in this part. Each correct answer will receive 4 credits. Clearly indicate the necessary steps, including appropriate formula substitutions, diagrams, graphs, charts, etc. For all questions in this part, a correct numerical answer with no work shown will receive only 1 credit.   [16 credits]

33 A school plans to have a fundraiser before basketball games selling shirts with their school logo. The school contacted two companies to find out how much it would cost to have the shirts made. Company A charges a $50 set-up fee and $5 per shirt. Company B charges a $25 set-up fee and $6 per shirt.

Write an equation for Company A that could be used to determine the total cost, A, when x shirts are ordered. Write a second equation for Company B that could be used to determine the total cost, B, when x shirts are ordered.

Determine algebraically and state the *minimum* number of shirts that must be ordered for it to be cheaper to use Company A.

34 Graph $y = f(x)$ and $y = g(x)$ on the set of axes below.

$$f(x) = 2x^2 - 8x + 3$$
$$g(x) = -2x + 3$$

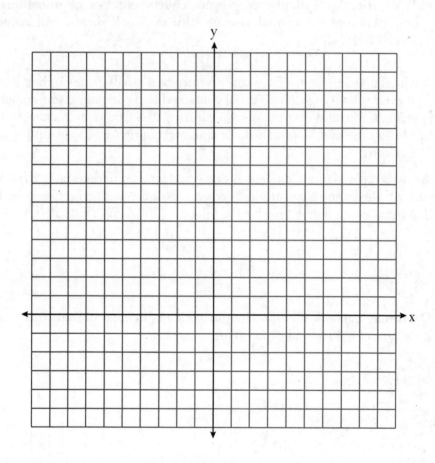

Determine and state all values of x for which $f(x) = g(x)$.

35 The table below shows the number of hours ten students spent studying for a
test and their scores.

Hours Spent Studying (x)	0	1	2	4	4	4	6	6	7	8
Test Scores (y)		35	40	46	65	67	70	82	88	82	95

Write the linear regression equation for this data set. Round all values to the
nearest hundredth.

State the correlation coefficient of this line, to the *nearest hundredth*.

Explain what the correlation coefficient suggests in the context of the
problem.

36 A system of inequalities is graphed on the set of axes below.

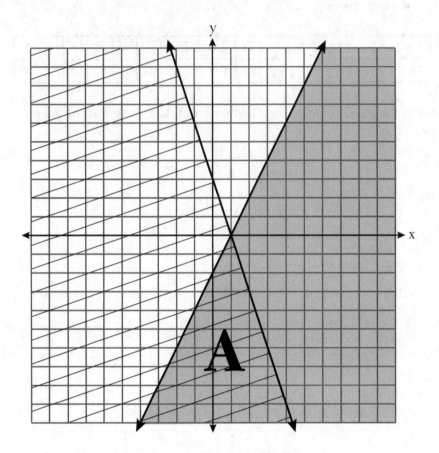

State the system of inequalities represented by the graph.

State what region *A* represents.

State what the entire gray region represents.

PART IV

Answer the question in this part. A correct answer will receive 6 credits. Clearly indicate the necessary steps, including appropriate formula substitutions, diagrams, graphs, charts, etc. A correct numerical answer with no work shown will receive only 1 credit.   [6 credits]

37 When visiting friends in a state that has no sales tax, two families went to a fast-food restaurant for lunch. The Browns bought 4 cheeseburgers and 3 medium fries for $16.53. The Greens bought 5 cheeseburgers and 4 medium fries for $21.11.

Using c for the cost of a cheeseburger and f for the cost of medium fries, write a system of equations that models this situation.

The Greens said that since their bill was $21.11, each cheeseburger must cost $2.49 and each order of medium fries must cost $2.87 each. Are they correct? Justify your answer.

Using your equations, algebraically determine both the cost of one cheeseburger and the cost of one order of medium fries.

Answers
June 2019
Algebra I

Answer Key

PART I

1. (4)	**5.** (1)	**9.** (4)	**13.** (4)	**17.** (3)	**21.** (2)
2. (2)	**6.** (1)	**10.** (1)	**14.** (3)	**18.** (3)	**22.** (1)
3. (4)	**7.** (1)	**11.** (3)	**15.** (2)	**19.** (2)	**23.** (2)
4. (2)	**8.** (3)	**12.** (2)	**16.** (2)	**20.** (4)	**24.** (1)

PART II

25. $x = 8$

26. $-3x^2 - 18x + 19$

27.

28. $\{-6, 6\}$

29. 1 a.m. to 12 noon because

$$2\frac{10}{11} > 2\frac{7}{9}$$

30. $h = \dfrac{3V}{\pi r^2}$

31. 8, 18, 38

32. $(1, -9)$

PART III

33. $A = 5x + 50, B = 6x + 25; x = 26$

34.

$x = 0, x = 3$

35. $y = 7.79x + 34.27, r = .98$

36. $y \leq 2x - 2, y < -3x + 3$; A represents the solution set of the system of inequalities. The gray region is the solution set to $y \leq 2x - 2$.

PART IV

37. $4c + 3f = 16.53, 5c + 4f = 21.11$. They are not correct because it would be \$23.93, not \$21.11; one cheeseburger is \$2.79, and one order of fries is \$1.79.

In **Parts II–IV**, you are required to show how you arrived at your answers. For sample methods of solutions, see the *Answers Explained* section.

Answers Explained

PART I

1. $w^4 - 36$ can be rewritten as $(w^2)^2 - 6^2$. This can be factored with the factoring pattern $a^2 - b^2 = (a - b)(a + b)$.

$$w^4 - 36 = (w^2)^2 - 6^2 = (w^2 - 6)(w^2 + 6)$$

The correct choice is (**4**).

2. $$f(-3) = 4(-3) + 5 = -12 + 5 = -7$$

The correct choice is (**2**).

3. Testing choice (1): When a relation is described in a table, it is only a function if there are no repeats in the x column. Since there are no repeats in this x column, this relation is a function.

Testing choice (2): When the relation is given as an equation, the simplest thing to do is to make a graph of that equation and see if it passes the vertical line test. The graph of a function will pass the "vertical line test," which means that there are no two points on the graph that have the same x-coordinate. If it is possible to draw a vertical line anywhere on the graph of a relation that will pass through more than one point, the relation will not be a function. The graph of the relation $3x + 2y = 4$ looks like this:

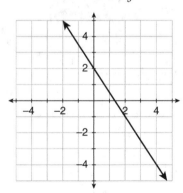

Since it passes the vertical line test, it is the graph of a function.

Testing choice (3): On this graph, there is no place where a vertical line will pass through more than one point, so this is the graph of a function.

Testing choice (4): When the relation is described with a mapping diagram, it is only a function if there is no place where the same element in the domain (the left oval) is being mapped to two different elements in the range (the right oval). Since in this mapping diagram the element 2 in the domain is mapped to both A and B, this relation is not a function.

The correct choice is (**4**).

4. Graph both of the functions on the same set of axes. The graph for $g(x)$ is 3 units to the right of the graph of $f(x)$.

For the TI-84:

For the TI-Nspire:

This question can also be done without a graphing calculator. An equation of the form $y = (x - h)^2 + k$ is called the *vertex form* of a parabola. The vertex will be (h, k).

The graph of function f will be a parabola with the vertex $(2, 4)$. The graph of the function g will be a parabola with the vertex $(5, 4)$. So, the graph of function g will be 3 units to the right of the graph of function f.

The correct choice is (**2**).

5. When a polynomial is written in standard form, the term with the largest exponent will be first and each term after that will have a smaller exponent than the previous term. The coefficients before the variables are not relevant to this problem.

For Anne, the sequence of exponents is 7, 5, 3, 1, so the exponents are in decreasing order.

For the other people, the exponents are not in decreasing order.
Bob: 3, 5, 7, 1
Carrie: 1, 7, 5, 3
Dylan: 1, 3, 5, 7

The correct choice is (**1**).

6. Plotting these four points on a graph looks like this:

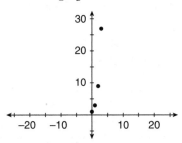

Here is what a graph looks like for each of the four choices:

Choice (1) exponential growth	Choice (2) exponential decay	Choice (3) linear with positive rate of change	Choice (4) linear with negative rate of change

Of the four choices, the one that the four plotted points most resembles is choice (1).

The correct choice is (**1**).

7. A rule of radicals is that $\sqrt{a} \cdot \sqrt{b} = \sqrt{ab}$. When this rule is applied to choice (1), it becomes $\sqrt{2} \cdot \sqrt{18} = \sqrt{2 \cdot 18} = \sqrt{36} = 6$, which is a rational number since it is a terminating decimal number. When an irrational number is expressed as a decimal, it will be nonterminating and nonrepeating.

The correct choice is (**1**).

8. The x-intercepts of this graph are $(-2, 0)$, $(1, 0)$, and $(2, 0)$. If $(a, 0)$ is an x-intercept of the graph of a polynomial function, one of its factors will be $(x - a)$. So, for this function, $f(x) = (x - (-2))(x - 1)(x - 2) = (x + 2)(x - 1)(x - 2)$. Though this is not one of the choices, if you multiply the first and third factors together, $(x + 2)(x - 2) = x^2 - 4$. So, the function can also be written as $f(x) = (x - 1)(x^2 - 4)$.

Another way you could approach this question would be to graph the answer choices to see which one matches the given graph. When choice (3) is graphed on the graphing calculator, it looks like this:

For the TI-84:

For the TI-Nspire:

The correct choice is (**3**).

9. She got from $p^2 + 5 = 8p - 7$ to $p^2 + 12 = 8p$ by adding 7 to both sides of the equation.

$$
\begin{array}{r}
p^2 + 5 = 8p - 7 \\
+\,7 = \quad\ +\,7 \\
\hline
p^2 + 12 = 8p
\end{array}
$$

The addition property of equality states that if you have an equality and you add the same number to both sides of the equal sign, the new equation is also true. This is what was done here.

The correct choice is (**4**).

10. The solution could be written as $x > 42$ and $x \leq 57$ since a person's height must meet both of those conditions. A shorthand for this "and" statement is $42 < x \leq 57$. This is <u>not</u> the same thing as choice (3), $42 < x$ or $x \leq 57$. The "or" statement is true if either of the conditions is met; so, for choice (3), someone under 42 inches would be allowed on the ride because his or her height is less than or equal to 57 inches.

The correct choice is (**1**).

11. Make a sketch of the scenario in each of the answer choices and see which one looks most like a line.

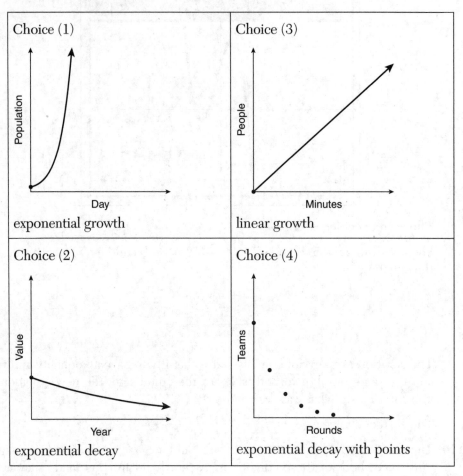

The correct choice is (**3**).

12. A total of $42 + 26 = 68$ people preferred burgers. Of those 68 people, 26 of them were female. So, the percentage of people who preferred burgers who were female is $\frac{26}{68} \approx .382 = 38.2\%$.

The correct choice is (**2**).

13. When the same number is added to or subtracted from both sides of an inequality, the inequality remains true. This inequality can be made to look like the answer choices by subtracting both $2a$ and $7b$ from both sides of the inequality.

$$\begin{array}{r} 3a + 7b > 2a - 8b \\ \underline{-2a - 7b = -2a - 7b} \\ a > -15b \end{array}$$

The correct choice is (**4**).

14. Finding the zeros, y-intercept, maximum, and minimum of a function is easiest to do with a graph. The zeros are the x-coordinates of the x-intercepts and the y-intercept is where the graph crosses the y-axis. Functions whose graphs are lines don't have a maximum or a minimum, but functions whose graphs are parabolas will either have a maximum or a minimum depending on whether the parabola looks more like an "n" or a "u."

The graph of the function for A looks like this:

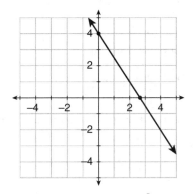

It has one zero at the x-intercept at $x = \dfrac{8}{3}$. The y-intercept is at $(0, 4)$. Because it is a line, it does not have a maximum or a minimum point.

The graph of the function for B looks like this:

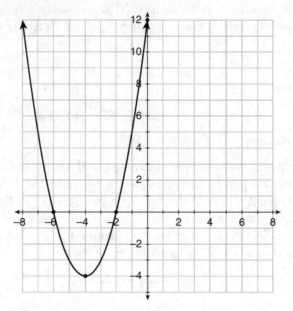

It has two zeros at the x-intercepts at $x = -2$ and $x = -6$. The y-intercept is at $(0, 12)$. Because it is a parabola shaped like a "u," it has a minimum point.

The graph of the function for C was given.

It has two zeros at the x-intercepts at $x = -2$ and $x = -4$. The y-intercept is at $(0, -8)$. Because it is a parabola shaped like an "n," it has a maximum point.

B has a minimum and C has a maximum.

The correct choice is (**3**).

15. If Nicci's age is a, her sister's age is $2a - 7$ (<u>not</u> $7 - 2a$). The sum of their ages is $a + (2a - 7)$. Since the sum of their ages is 41, the equation is $a + (2a - 7) = 41$.

The correct choice is (**2**).

16. Since the population is increasing each year, this is an example of exponential growth. For an equation of the form $y = ab^x$, exponential growth occurs when $b > 1$. This means that Roman numeral IV must be part of the solution.

Since in 2013 the population is 3810, $P(0) = 1310$. This fact can be used to solve for a.

$P(0) = ab^0 = 3810$

$P(0) = a \cdot 1 = 3810$

$a = 3810$

This means that Roman numeral I must be part of the solution.

In general, in an equation of the form $y = ab^x$, the y-intercept will be a value of a.

The correct choice is (**2**).

17. The simplest way to solve this is to multiply all the answer choices to see which becomes $4w^2 - 11w - 3$.

Testing choice (1): $(2w + 1)(2w - 3) = 4w^2 - 6w + 2w - 3 = 4w^2 - 5w - 3$
Testing choice (2): $(2w - 1)(2w + 3) = 4w^2 + 6w - 2w - 3 = 4w^2 + 5w - 3$
Testing choice (3): $(4w + 1)(w - 3) = 4w^2 - 12w + 1w - 3 = 4w^2 - 11w - 3$
Testing choice (4): $(4w - 1)(w + 3) = 4w^2 + 12w - 1w - 3 = 4w^2 + 11w - 3$

The correct choice is (**3**).

18. For each answer choice, replace the x in the equation with the x-coordinate of the ordered pair and y with the y-coordinate of the ordered pair, and see which one does not result in the left side of the equation equaling the right side of the equation.

Testing choice (1):

$15.25 \stackrel{?}{=} 3(-1.5)^2 - (-1.5) + 7$
$15.25 \stackrel{?}{=} 3 \cdot 2.25 + 1.5 + 7$
$15.25 \stackrel{\checkmark}{=} 15.25$

Testing choice (2):

$7.25 \stackrel{?}{=} 3 \cdot 5^2 - 0.5 + 7$
$7.25 \stackrel{?}{=} 3 \cdot 25 - 0.5 + 7$
$7.25 \stackrel{\checkmark}{=} 7.25$

Testing choice (3):

$10.25 \stackrel{?}{=} 3 \cdot 1.25^2 - 1.25 + 7$
$10.25 \stackrel{?}{=} 3 \cdot 1.5625 - 1.25 + 7$
$10.25 \neq 10.4375$

Testing choice (4):

$23.25 \stackrel{?}{=} 3 \cdot 2.5^2 - 2.5 + 7$
$23.25 \stackrel{?}{=} 3 \cdot 6.25 - 2.5 + 7$
$23.25 \stackrel{\checkmark}{=} 23.25$

The correct choice is (**3**).

19. An arithmetic sequence is one in which each term is equal to the previous term plus or minus some constant.

    Testing Roman numeral I: Since $4 = 2 + 2$, $6 = 4 + 2$, $8 = 6 + 2$, etc., this is an arithmetic series with a first term of 2 and a common difference of 2.

    Testing Roman numeral II: Since $4 = 2 + 2$, $8 = 4 \cdot 2$, $16 = 8 \cdot 2$, etc., this is <u>not</u> an arithmetic series. It is a geometric series.

    Testing Roman numeral III: Since $a + 2 = a + 2$, $a + 4 = a + 2 + 2$, $a + 6 = a + 4 + 2$, etc., this is an arithmetic series with a first term of a and a common difference of 2.

    The correct choice is (**2**).

20. Choices (1) and (2) include negative numbers. Since you cannot purchase a negative amount of beef, you can eliminate these choices. Since beef does not have to be purchased in one-pound units (you can buy, for example, 1.7 pounds of beef), choice (3) can be eliminated also.

    The correct choice is (**4**).

21. Using the quadratic formula with $a = 1$, $b = -5$, and $c = -4$,

$$x = \frac{-b \pm \sqrt{b^2 - 4ac}}{2a} = \frac{-(-5) \pm \sqrt{(-5)^2 - 4(1)(-4)}}{2(1)}$$

$$= \frac{5 \pm \sqrt{25 + 16}}{2} = \frac{5 \pm \sqrt{41}}{2}$$

    The correct choice is (**2**).

22. Population standard deviation can be calculated with a graphing calculator.

    For the TI-84:

    Press [STAT] [1]. Clear all lists and enter the thirteen numbers into L1.

Press [STAT], [Right Arrow], [1] and select Calculate.

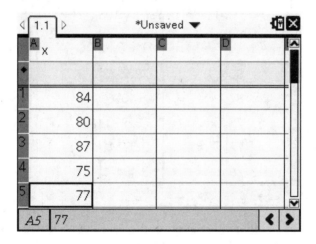

For the TI-NSpire:

From the home screen, select the Add Lists & Spreadsheet icon. Name column A x and fill in cells A1 to A13 with the thirteen numbers.

Press [menu] [4] [1] [1] for One-Variable Statistics.

Select OK twice.

The population standard deviation is the one denoted with the symbol σx. Rounded to the nearest tenth, it is 3.5.

The correct choice is (**1**).

23. In an exponential equation of the form $y = a(1 - r)^x$, r is the decay rate. For this question $1 - r = 0.98$, so $r = 0.02$.

    The correct choice is (**2**).

24. From the reference sheet provided on the Regents, 1 inch = 2.54 centimeters. To convert 91 centimeters into inches, divide 91 by 2.54 to get approximately 35.8 inches. Since 1 day equals 24 hours, the growth rate in inches per hour can be calculated by dividing the total growth by the total number of hours.

$$\frac{35.8}{24} \approx 1.49 \text{ inches per hour}$$

    The correct choice is (**1**).

PART II

25.

$$-\frac{2}{3}(x + 12) + \frac{2}{3}x = -\frac{5}{4}x + 2$$

$$-\frac{2}{3}x - 8 + \frac{2}{3}x = -\frac{5}{4}x + 2$$

$$-8 = -\frac{5}{4}x + 2$$

$$-2 = -2$$

$$-10 = -\frac{5}{4}x$$

$$-\frac{4}{5}(-10) = -\frac{4}{5}\left(-\frac{5}{4}x\right)$$

$$8 = x$$

26. $C = G - 3F = 3x^2 + 4 - 3(2x^2 + 6x - 5) = 3x^2 + 4 - 6x^2 - 18x + 15$
$= -3x^2 - 18x + 19$

27. A piecewise function is one that uses more than one rule, depending on what the input value to the function is. Make a table of values from $x = -5$ to $x = 6$. For the first seven values, use the top part of the piecewise function. For the last five values, use the bottom part of the piecewise function.

x	$f(x)$
-5	$|-5| = 5$
-4	$|-4| = 4$
-3	$|-3| = 3$
-2	$|-2| = 2$
-1	$|-1| = 1$
0	$|0| = 0$
1	$|1| = 1$
2	$-2(2) + 10 = -4 + 10 = 6$
3	$-2(3) + 10 = -6 + 10 = 4$
4	$-2(4) + 10 = -8 + 10 = 2$
5	$-2(5) + 10 = -10 + 10 = 0$
6	$-2(6) + 10 = -12 + 10 = -2$

Plot these twelve points.

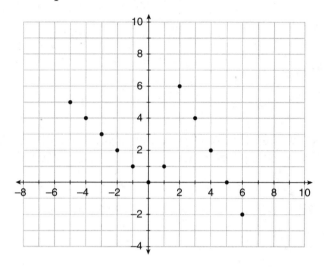

For $x = 2$, the bottom part of the piecewise function was used. For graphing, it is also useful to substitute $x = 2$ into the top part of the piecewise function because it is the boundary between the two pieces. When $x = 2$ is substituted into the top part of the piecewise function, the expression evaluates to 2. Plot the point $(2, 2)$ as an open circle since it is not an actual point on the graph. Then, connect the dots like this:

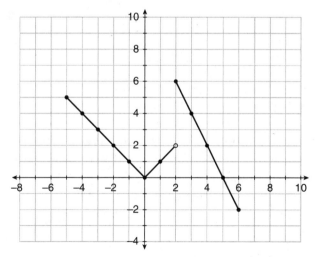

28.
$$\frac{5x^2}{5} = \frac{180}{5}$$
$$x^2 = 36$$
$$\sqrt{x^2} = \pm\sqrt{36}$$
$$x = \pm 6$$
$$\{-6, 6\}$$

29. To calculate the rate of snowfall for an interval of time, divide the amount of snow that fell in that interval by the amount of time that elapsed in that interval.

For the 1 a.m. to 12 noon interval, the rate of snowfall is $\frac{33-1}{12-1} = \frac{32}{11} = 2\frac{10}{11}$ inches per hour.

For the 6 a.m. to 3 p.m. interval, the rate of snowfall is $\frac{36-11}{15-6} = \frac{25}{9} = 2\frac{7}{9}$ inches per hour.

Since $2\frac{10}{11} > 2\frac{7}{9}$, the 1 a.m. to 12 noon rate of snowfall is greater than the 6 a.m. to 3 p.m. rate of snowfall.

30.
$$V = \frac{1}{3}\pi r^2 h$$
$$3V = 3 \cdot \frac{1}{3}\pi r^2 h$$
$$3V = \pi r^2 h$$
$$\frac{3V}{\pi r^2} = \frac{\pi r^2 h}{\pi r^2}$$
$$\frac{3V}{\pi r^2} = h$$

31. A recursive formula describes how to get the next term of a sequence based on the previous terms. Substituting $n = 2, 3,$ and 4 into the recursive part of the equation, it becomes
$$a_2 = 2(a_{2-1} + 1) = 2(a_1 + 1) = 2(3 + 1) = 2 \cdot 4 = 8$$
$$a_3 = 2(a_{3-1} + 1) = 2(a_2 + 1) = 2(8 + 1) = 2 \cdot 9 = 18$$
$$a_4 = 2(a_{4-1} + 1) = 2(a_3 + 1) = 2(18 + 1) = 2 \cdot 19 = 38$$

32. The vertex form of a quadratic equation is $y = a(x - h)^2 + k$, where the vertex is (h, k). Completing the square is when you create a quadratic trinomial where the constant term is the square of half the coefficient on the x.

In the function $f(x) = x^2 - 2x - 8$, the coefficient on the x is -2. Calculate the square of half of this coefficient $\left(\dfrac{-2}{2}\right)^2 = (-1)^2 = +1$.

To complete the square, rewrite the -8 as $+1 - 9$. The function becomes $f(x) = x^2 - 2x + 1 - 9$. The first three terms are now a perfect square trinomial and can be factored.

$$f(x) = x^2 - 2x + 1 - 9 = (x - 1)^2 - 9$$

This is now in vertex form and the vertex is $(1, -9)$.

PART III

33. For company A, x shirts cost \$$5x$ plus the \$50 set-up fee. The equation is $A = 5x + 50$. For company B, x shirts cost \$$6x$ plus the \$25 set-up fee. The equation is $B = 6x + 25$.

For some values of x, A is more expensive and, for others, B is more expensive. To find the minimum number of shirts for it to be cheaper to use Company A, find the smallest integer that makes the inequality $5x + 50 < 6x + 25$ true.

Solving the inequality becomes

$$5x + 50 < 6x + 25$$
$$-6x = -6x$$
$$-x + 50 < 25$$
$$-50 = -50$$
$$-x < -25$$
$$\frac{-x}{-1} > \frac{-25}{-1}$$
$$x > 25$$

The smallest integer that makes this true is 26. 25 is not the solution since it is not cheaper to buy 25 shirts from company A; it is the same price as company B.

34. The graph of function f is a parabola, and the graph of function g is a line. An important point to graph for a parabola is the vertex. In general, the x-coordinate of the vertex of a parabola is $x = -\dfrac{b}{2a} = -\dfrac{-8}{2 \cdot 2} = 2$. Choose at least three values smaller than 2 and three values greater than 2 for the x values in your chart.

x	$f(x)$	$g(x)$
-1	$2(-1)^2 - 8(-1) + 3 = 2 + 8 + 3 = 13$	$-2(-1) + 3 = 2 + 3 = 5$
0	$2(0)^2 - 8(0) + 3 = 0 + 0 + 3 = 3$	$-2(0) + 3 = 0 + 3 = 3$
1	$2(1)^2 - 8(1) + 3 = 2 - 8 + 3 = -3$	$-2(1) + 3 = -2 + 3 = 1$
2	$2(2)^2 - 8(2) + 3 = 8 - 16 + 3 = -5$	$-2(2) + 3 = -4 + 3 = -1$
3	$2(3)^2 - 8(3) + 3 = 18 - 24 + 3 = -3$	$-2(3) + 3 = -6 + 3 = -3$
4	$2(4)^2 - 8(4) + 3 = 32 - 32 + 3 = 3$	$-2(4) + 3 = -8 + 3 = -5$
5	$2(5)^2 - 8(5) + 3 = 50 - 40 + 3 = 13$	$-2(5) + 3 = -10 + 3 = -7$

The graph looks like this:

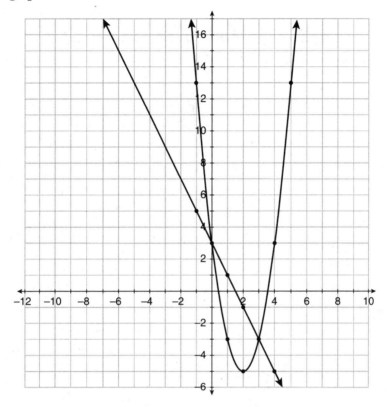

According to the chart, $f(0) = g(0) = 3$ and $f(3) = g(3) = -3$, so the two values of x for which $f(x) = g(x)$ are 0 and 3.

It can also be seen on the graph that the x-coordinates of the two points where the line intersects the parabola are 0 and 3.

These functions can also be graphed on a graphing calculator.

For the TI-84:

For the TI-Nspire:

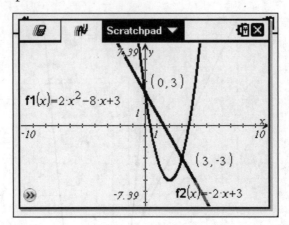

35. A line of best fit for a data set can be calculated with a graphing calculator.

    For the TI-84:

    To turn diagnostics on, press [2nd] [0], scroll down to select DiagnosticOn, and press [ENTER].

Press [STAT] [1] and enter the *x* values into L1 and the *y* values into L2.

L1	L2	L3	2
4	65		
4	67		
4	70		
6	82		
6	88		
7	82		
8	95		

L2(10) =95

Press [STAT] [Right Arrow] [4] to get the LinReg function. Select Calculate.

LinReg
y=ax+b
a=7.792207792
b=34.27272727
r²=.9634878259
r=.9815741571

For the TI-NSpire:

From the home screen select the Add Lists & Spreadsheet icon. In the first row, label the A column "hours" and the B column "score." Enter the hours in cells A1 to A9. Enter the scores in cells B1 to B9.

	A hours	B score	C	D	
◆					
1	0	35			
2	1	40			
3	2	46			
4	4	65			
5	4	67			

Move the cursor to cell C1. Press [menu] [4] [1] [3]. Set X List to "hours" and Y List to "score," and select OK.

The line of best fit is $y = 7.79x + 34.27$.

This is the line of best fit graphed together with the scatter plot.

For the TI-84:

For the TI-Nspire:

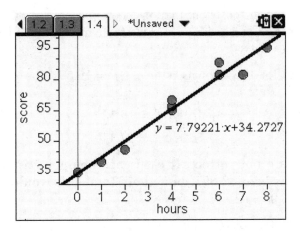

The correlation coefficient (r) is a measure of how close the points are to the line of best fit. If the line of best fit has a positive slope, the correlation coefficient will be positive. If the line of best fit has a negative slope, the correlation coefficient will be negative. The closer the correlation coefficient is to 1 (or to -1), the closer to the line of best fit the points will lie.

For this data set, the correlation coefficient is .98, which is very close to 1, so the data set is very nearly a line.

36. The solid line has a y-intercept of $(0, -2)$ and, since it passes through $(1, 0)$, it has a slope of 2. The equation of the solid line is $y = 2x - 2$. Since the area underneath that line is shaded, the first inequality is $y \leq 2x - 2$.

The dotted line has a y-intercept of $(0, 3)$ and, since it passes through $(1, 0)$, it has a slope of -3. The equation of the dotted line is $y = -3x + 3$. Since the area underneath that line is shaded, the second inequality is $y < -3x + 3$. There is a "$<$" instead of a "$\leq$."

So, the system of inequalities is

$$y \leq 2x - 2$$
$$y < -3x - 3$$

The region A represents all the ordered pairs that make up the solution set to the system of inequalities. Ordered pairs in this region will make both of the inequalities true.

The shaded gray region represents all the ordered pairs that make up the solution set to the inequality $y \leq 2x - 2$.

PART IV

37. If one cheeseburger costs $\$c$, then 4 cheeseburgers are $\$4c$ and 5 cheeseburgers are $\$5c$. If one order of fries is $\$f$, then 3 orders of fries are $\$3f$ and 4 orders of fries are $\$4f$.

The equation for the Browns is $4c + 3f = 16.53$. The equation for the Greens is

$$4c + 3f = 16.53$$
$$5c + 4f = 21.11$$

The Greens are not correct. If each cheeseburger was \$2.49 and each order of fries was \$2.87, then their total bill would be $5 \times 2.49 + 4 \times 2.87 = 23.93 \neq 21.11$.

To solve the system of equations, multiply both sides of the top equation by -4 and both sides of the bottom equation by 3. This makes the coefficients of the f variables opposites; so, when the two equations are added together, the f will be eliminated.

$$-16c - 12f = -66.12$$
$$15c + 12f = 63.33$$

Adding the two equations together becomes

$$-c = -2.79$$
$$c = 2.79$$

The price of one cheeseburger is \$2.79.

To find the price of an order of fries, substitute $c = 2.79$ into either of the original equations and solve for f.

$$4(2.79) + 3f = 16.53$$
$$11.16 + 3f = 16.53$$
$$-11.16 = -11.16$$
$$\frac{3f}{3} = \frac{5.37}{3}$$
$$f = 1.79$$

The price of one order of fries is \$1.79.

Topic	Question Numbers	Number of Points	Your Points	Your Percentage
1. Polynomials	5, 8, 26, 32	$2 + 2 + 2 + 2 = 8$		
2. Properties of Algebra	9, 25, 28, 30	$2 + 2 + 2 + 2 = 8$		
3. Functions	2, 3, 20, 29	$2 + 2 + 2 + 2 = 8$		
4. Creating and Interpreting Equations	11, 15, 23	$2 + 2 + 2 = 6$		
5. Inequalities	10, 13, 36	$2 + 2 + 4 = 8$		
6. Sequences and Series	19, 31	$2 + 2 = 4$		
7. Systems of Equations	33, 37	$4 + 6 = 10$		
8. Quadratic Equations and Factoring	1, 17, 21	$2 + 2 + 2 = 6$		
9. Regression	35	4		
10. Exponential Equations	6, 16	$2 + 2 = 4$		
11. Graphing	4, 14, 18, 27, 34	$2 + 2 + 2 + 2 + 4 = 12$		
12. Statistics	12, 22	$2 + 2 = 4$		
13. Number Properties	7	2		
14. Unit Conversions	24	2		

HOW TO CONVERT YOUR RAW SCORE TO YOUR ALGEBRA I REGENTS EXAMINATION SCORE

The accompanying conversion chart must be used to determine your final score on the June 2019 Regents Examination in Algebra I. To find your final exam score, locate in the column labeled "Raw Score" the total number of points you scored out of a possible 86 points. Since partial credit is allowed in Parts II, III, and IV of the test, you may need to approximate the credit you would receive for a solution that is not completely correct. Then locate in the adjacent column to the right the scale score that corresponds to your raw score. The scale score is your final Algebra I Regents Examination score.

Regents Examination in Algebra I—June 2019
Chart for Converting Total Test Raw Scores to Final
Exam Scores (Scale Scores)

Raw Score	Scale Score	Performance Level	Raw Score	Scale Score	Performance Level	Raw Score	Scale Score	Performance Level
86	100	5	57	81	4	28	66	3
85	99	5	56	81	4	27	65	3
84	97	5	55	81	4	26	64	2
83	96	5	54	80	4	25	63	2
82	95	5	53	80	4	24	61	2
81	94	5	52	80	4	23	60	2
80	93	5	51	80	4	22	58	2
79	92	5	50	79	3	21	57	2
78	91	5	49	79	3	20	55	2
77	90	5	48	79	3	19	54	1
76	90	5	47	78	3	18	52	1
75	89	5	46	78	3	17	50	1
74	88	5	45	78	3	16	48	1
73	88	5	44	77	3	15	46	1
72	87	5	43	77	3	14	44	1
71	86	5	42	76	3	13	42	1
70	86	5	41	76	3	12	39	1
69	86	5	40	75	3	11	37	1
68	85	5	39	75	3	10	34	1
67	84	4	38	74	3	9	31	1
66	84	4	37	74	3	8	28	1
65	84	4	36	73	3	7	25	1
64	83	4	35	72	3	6	22	1
63	83	4	34	72	3	5	19	1
62	83	4	33	71	3	4	15	1
61	82	4	32	70	3	3	12	1
60	82	4	31	69	3	2	8	1
59	82	4	30	68	3	1	4	1
58	82	4	29	67	3	0	0	1

Examination
August 2019
Algebra I

HIGH SCHOOL MATH REFERENCE SHEET

Conversions

1 inch = 2.54 centimeters

1 meter = 39.37 inches

1 mile = 5280 feet

1 mile = 1760 yards

1 mile = 1.609 kilometers

1 kilometer = 0.62 mile

1 pound = 16 ounces

1 pound = 0.454 kilogram

1 kilogram = 2.2 pounds

1 ton = 2000 pounds

1 cup = 8 fluid ounces

1 pint = 2 cups

1 quart = 2 pints

1 gallon = 4 quarts

1 gallon = 3.785 liters

1 liter = 0.264 gallon

1 liter = 1000 cubic centimeters

Formulas

Triangle	$A = \frac{1}{2}bh$
Parallelogram	$A = bh$
Circle	$A = \pi r^2$
Circle	$C = \pi d$ or $C = 2\pi r$

Formulas (continued)

General Prisms	$V = Bh$
Cylinder	$V = \pi r^2 h$
Sphere	$V = \dfrac{4}{3}\pi r^3$
Cone	$V = \dfrac{1}{3}\pi r^2 h$
Pyramid	$V = \dfrac{1}{3}Bh$
Pythagorean Theorem	$a^2 + b^2 = c^2$
Quadratic Formula	$x = \dfrac{-b \pm \sqrt{b^2 - 4ac}}{2a}$
Arithmetic Sequence	$a_n = a_1 + (n-1)d$
Geometric Sequence	$a_n = a_1 r^{n-1}$
Geometric Series	$S_n = \dfrac{a_1 - a_1 r^n}{1 - r}$ where $r \neq 1$
Radians	$1 \text{ radian} = \dfrac{180}{\pi} \text{ degrees}$
Degrees	$1 \text{ degree} = \dfrac{\pi}{180} \text{ radians}$
Exponential Growth/Decay	$A = A_0 e^{k(t-t_0)} + B_0$

PART I

Answer all 24 questions in this part. Each correct answer will receive 2 credits. No partial credit will be allowed. For each statement or question, write in the space provided the numeral preceding the word or expression that best completes the statement or answers the question. [48 credits]

1 Bryan's hockey team is purchasing jerseys. The company charges $250 for a onetime set-up fee and $23 for each printed jersey. Which expression represents the total cost of x number of jerseys for the team?

(1) $23x$ (3) $23x + 250$

(2) $23 + 250x$ (4) $23(x + 250)$ 1 _____

2 Which table represents a function?

x	y
2	-3
3	0
4	-3
2	1

(1)

x	y
-3	0
-2	1
-3	2
2	3

(3)

x	y
1	2
1	3
1	4
1	5

(2)

x	y
-2	-4
0	2
2	4
4	6

(4)   2 _____

3 Which expression is equivalent to $2(x^2 - 1) + 3x(x - 4)$?

(1) $5x^2 - 5$ (3) $5x^2 - 12x - 1$

(2) $5x^2 - 6$ (4) $5x^2 - 12x - 2$ 3 ____

4 The value of x that satisfies the equation $\dfrac{4}{3} = \dfrac{x + 10}{15}$ is

(1) -6 (3) 10

(2) 5   (4) 30   4 ____

5 Josh graphed the function $f(x) = -3(x - 1)^2 + 2$. He then graphed the function $g(x) = -3(x - 1)^2 - 5$ on the same coordinate plane. The vertex of $g(x)$ is

(1) 7 units below the vertex of $f(x)$

(2) 7 units above the vertex of $f(x)$

(3) 7 units to the right of the vertex of $f(x)$

(4) 7 units to the left of the vertex of $f(x)$ 5 ____

6 A survey was given to 12th-grade students of West High School to determine the location for the senior class trip. The results are shown in the table below.

	Niagara Falls	Darien Lake	New York City
Boys	56	74	103
Girls	71	92	88

To the *nearest percent*, what percent of the boys chose Niagara Falls?

(1) 12   (3) 44

(2) 24   (4) 56   6 ____

7 Which type of function is shown in the graph below?

(1) linear   (3) square root

(2) exponential   (4) absolute value   7 _____

8 The expression $16x^2 - 81$ is equivalent to

(1) $(8x - 9)(8x + 9)$ (3) $(4x - 9)(4x + 9)$

(2) $(8x - 9)(8x - 9)$ (4) $(4x - 9)(4x - 9)$ 8 _____

9 The owner of a landscaping business wants to know how much time, on average, his workers spend mowing one lawn. Which is the most appropriate rate with which to calculate an answer to his question?

(1) lawns per employee   (3) employee per lawns

(2) lawns per day   (4) hours per lawn   9 _____

10 A ball is thrown into the air from the top of a building. The height, $h(t)$, of the ball above the ground t seconds after it is thrown can be modeled by $h(t) = -16t^2 + 64t + 80$. How many seconds after being thrown will the ball hit the ground?

(1) 5   (3) 80

(2) 2   (4) 144   10 _____

11 Which equation is equivalent to $y = x^2 + 24x - 18$?

(1) $y = (x + 12)^2 - 162$ (3) $y = (x - 12)^2 - 162$
(2) $y = (x + 12)^2 + 126$ (4) $y = (x - 12)^2 + 126$ 11 _____

12 When $(x)(x - 5)(2x + 3)$ is expressed as a polynomial in standard
form, which statement about the resulting polynomial is true?

(1) The constant term is 2.
(2) The leading coefficient is 2.
(3) The degree is 2.
(4) The number of terms is 2.   12 _____

13 The population of a city can be modeled by $P(t) = 3810(1.0005)^{7t}$,
where $P(t)$ is the population after t years. Which function is
approximately equivalent to $P(t)$?

(1) $P(t) = 3810(0.1427)^t$ (3) $P(t) = 26,670(0.1427)^t$
(2) $P(t) = 3810(1.0035)^t$ (4) $P(t) = 26,670(1.0035)^t$ 13 _____

14 The functions $f(x)$ and $g(x)$ are graphed on the set of axes below.

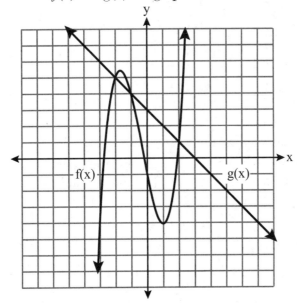

For which value of x is $f(x) \neq g(x)$?

(1) −1   (3) 3
(2) 2   (4) −2   14 _____

15 What is the range of the box plot shown below?

(1) 7   (3) 3

(2) 2   (4) 4   15 _____

16 Which expression is *not* equivalent to $2x^2 + 10x + 12$?

(1) $(2x + 4)(x + 3)$ (3) $(2x + 3)(x + 4)$

(2) $(2x + 6)(x + 2)$ (4) $2(x + 3)(x + 2)$ 16 _____

17 The quadratic functions $r(x)$ and $q(x)$ are given below.

x	r(x)
−4	−12
−3	−15
−2	−16
−1	−15
0	−12
1	7

$q(x) = x^2 + 2x - 8$

The function with the *smallest* minimum value is

(1) $q(x)$, and the value is −9   (3) $r(x)$, and the value is −16

(2) $q(x)$, and the value is −1   (4) $r(x)$, and the value is −2   17 _____

18 A child is playing outside. The graph below shows the child's distance, $d(t)$, in yards from home over a period of time, t, in seconds.

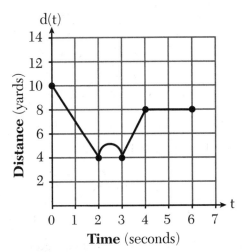

Which interval represents the child constantly moving closer to home?

(1) $0 \leq t \leq 2$ (3) $3 \leq t \leq 4$
(2) $2 \leq t \leq 3$ (4) $4 \leq t \leq 6$ 18 _____

19 If $a_1 = 6$ and $a_n = 3 + 2(a_{n-1})^2$, then a_2 equals

(1) 75   (3) 180
(2) 147   (4) 900   19 _____

20 The length of a rectangular patio is 7 feet more than its width, w. The area of a patio, $A(w)$, can be represented by the function

(1) $A(w) = w + 7$ (3) $A(w) = 4w + 14$
(2) $A(w) = w^2 + 7w$ (4) $A(w) = 4w^2 + 28w$ 20 _____

21 A dolphin jumps out of the water and then back into the water. His jump could be graphed on a set of axes where x represents time and y represents distance above or below sea level. The domain for this graph is best represented using a set of

(1) integers   (3) real numbers
(2) positive integers   (4) positive real numbers   21 _____

22 Which system of linear equations has the same solution as the one shown below?

$$x - 4y = -10$$
$$x + y = 5$$

(1) $5x = 10$
$x + y = 5$

(2) $-5y = -5$
$x + y = 5$

(3) $-3x = -30$
$x + y = 5$

(4) $-5y = -5$
$x - 4y = -10$

22 _____

23 Which interval represents the range of the function $h(x) = 2x^2 - 2x - 4$?

(1) $(0.5, \infty)$

(2) $(-4.5, \infty)$

(3) $[0.5, \infty)$

(4) $[-4.5, \infty)$

23 _____

24 What is a common ratio of the geometric sequence whose first term is 5 and third term is 245?

(1) 7

(2) 49

(3) 120

(4) 240

24 _____

PART II

Answer all 8 questions in this part. Each correct answer will receive 2 credits. Clearly indicate the necessary steps, including appropriate formula substitutions, diagrams, graphs, charts, etc. For all questions in this part, a correct numerical answer with no work shown will receive only 1 credit. [16 credits]

25 If $g(x) = -4x^2 - 3x + 2$, determine $g(-2)$.

26 A student is in the process of solving an equation. The original equation and the first step are shown below.

Original: $3a + 6 = 2 - 5a + 7$
Step one: $3a + 6 = 2 + 7 - 5a$

Which property did the student use for the first step? Explain why this property is correct.

27 On the set of axes below, graph the line whose equation is
$2y = -3x - 2$.

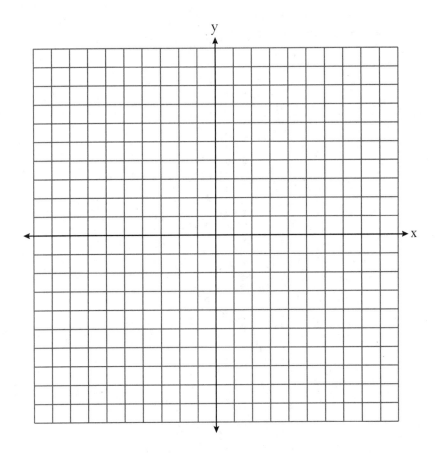

This linear equation contains the point $(2,k)$. State the value of k.

28 The formula $a = \dfrac{v_f - v_i}{t}$ is used to calculate acceleration as the change in velocity over the period of time.

Solve the formula for the final velocity, v_f, in terms of initial velocity, v_i, acceleration, a, and time, t.

29 Solve $\frac{3}{5}x + \frac{1}{3} < \frac{4}{5}x - \frac{1}{3}$ for x.

30 Is the product of two irrational numbers always irrational? Justify your answer.

31 Solve $6x^2 - 42 = 0$ for the exact values of x.

32 Graph the function: $h(x) = \begin{cases} 2x - 3, & x < 0 \\ x^2 - 4x - 5, & 0 \le x \le 5 \end{cases}$

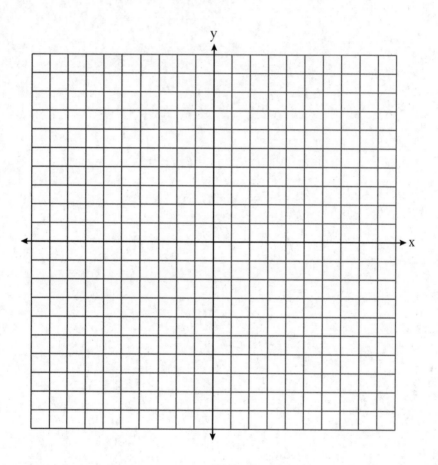

PART III

Answer all 4 questions in this part. Each correct answer will receive 4 credits. Clearly indicate the necessary steps, including appropriate formula substitutions, diagrams, graphs, charts, etc. For all questions in this part, a correct numerical answer with no work shown will receive only 1 credit.   [16 credits]

33 On the set of axes below, graph the following system of inequalities:

$$2x + y \geq 8$$
$$y - 5 < 3x$$

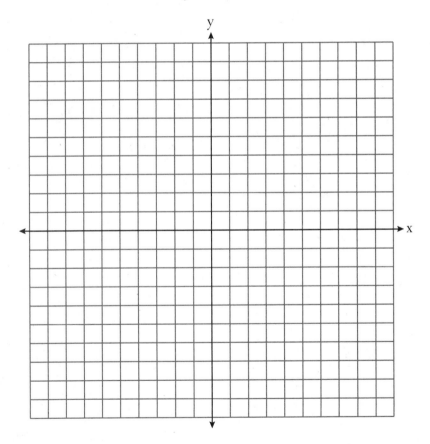

Determine if the point $(1,8)$ is in the solution set. Explain your answer.

34 On the day Alexander was born, his father invested $5000 in an account with a 1.2% annual growth rate. Write a function, $A(t)$, that represents the value of this investment t years after Alexander's birth.

Determine, to the *nearest dollar*, how much more the investment will be worth when Alexander turns 32 than when he turns 17.

35 Stephen collected data from a travel website. The data included a hotel's distance from Times Square in Manhattan and the cost of a room for one weekend night in August. A table containing these data appears below.

Distance From Times Square (city blocks) (x)	0	0	1	1	3	4	7	11	14	19
Cost of a Room (dollars) (y)	293	263	244	224	185	170	219	153	136	111

Write the linear regression equation for this data set. Round all values to the *nearest hundredth*.

State the correlation coefficient for this data set, to the *nearest hundredth*.

Explain what the sign of the correlation coefficient suggests in the context of the problem.

36 A snowstorm started at midnight. For the first 4 hours, it snowed at an average rate of one-half inch per hour.

The snow then started to fall at an average rate of one inch per hour for the next 6 hours.

Then it stopped snowing for 3 hours.

Then it started snowing again at an average rate of one-half inch per hour for the next 4 hours until the storm was over.

On the set of axes below, graph the amount of snow accumulated over the time interval of the storm.

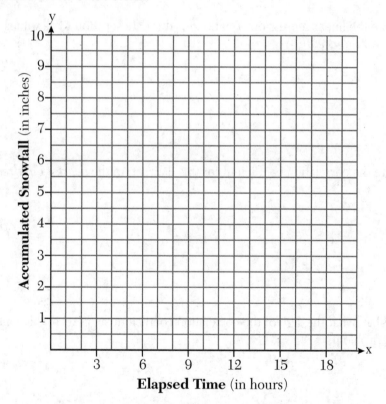

Determine the average rate of snowfall over the length of the storm. State the rate, to the *nearest hundredth of an inch per hour*.

PART IV

Answer the question in this part. A correct answer will receive 6 credits. Clearly indicate the necessary steps, including appropriate formula substitutions, diagrams, graphs, charts, etc. A correct numerical answer with no work shown will receive only 1 credit.   [6 credits]

37 Allysa spent $35 to purchase 12 chickens. She bought two different types of chickens. Americana chickens cost $3.75 each and Delaware chickens cost $2.50 each.

   Write a system of equations that can be used to determine the number of Americana chickens, A, and the number of Delaware chickens, D, she purchased.

   Determine algebraically how many of each type of chicken Allysa purchased.

   Each Americana chicken lays 2 eggs per day and each Delaware chicken lays 1 egg per day. Allysa only sells eggs by the full dozen for $2.50. Determine how much money she expects to take in at the end of the first week with her 12 chickens.

Answers
August 2019
Algebra I

Answer Key

PART I

1. (3)	**5.** (1)	**9.** (4)	**13.** (2)	**17.** (3)	**21.** (4)
2. (4)	**6.** (2)	**10.** (1)	**14.** (3)	**18.** (1)	**22.** (1)
3. (4)	**7.** (2)	**11.** (1)	**15.** (1)	**19.** (1)	**23.** (4)
4. (3)	**8.** (3)	**12.** (2)	**16.** (3)	**20.** (2)	**24.** (1)

PART II

25. -8

26. Commutative property of addition

27. $k = -4$

28. $v_f = at + v_i$

29. $x > \dfrac{10}{3}$

30. No. For example, $\sqrt{2} \cdot \sqrt{2} = 2$.

31. $x = \pm\sqrt{7}$

32.

PART III

33. (1, 8) is not part of the solution set.

34. $A(t) = 5000(1.012)^t$, \$1200

35. $y = -7.76x + 246.34$, $r = -0.88$, which is negative because the line of best fit has a negative slope.

36. 0.59 inches per hour

PART IV

37. $A + D = 12$
$3.75A + 2.50D = 35$
$A = 4, D = 8$, \$22.50

In **Parts II–IV**, you are required to show how you arrived at your answers. For sample methods of solutions, see the *Answers Explained* section.

Answers Explained

PART I

1. Since each printed jersey costs 23 dollars, x jerseys cost $23x$ dollars. When the cost of the jerseys is added to the one-time setup fee of 250 dollars, the total cost is then $23x + 250$ dollars.

   In general, when an equation of the form $mx + b$ models a real-world scenario, the b represents the fixed cost and the m represents the increase in the total for each time x increases by 1.

   The correct choice is **(3)**.

2. When a function is represented in a table, there will be no repeated values in the x column. In choice (1), there are two 2's. In choice (2), there are four 1's. In choice (3), there are two −3's. In choice (4), there are no repeated values in the x column, so this table represents a function.

   Even if in choice (4) there were repeated values among the y-values, it would still be a function. All that matters is that there are no repeated values among the x-values.

   The correct choice is **(4)**.

3. Use the distributive property and then combine like terms.

   $$2\left(x^2 - 1\right) + 3x(x - 4) = 2x^2 - 2 + 3x^2 - 12x$$
   $$= 5x^2 - 12x - 2$$

   The correct choice is **(4)**.

4. Cross multiply and then solve for x.

   $$\frac{4}{3} = \frac{x + 10}{15}$$
   $$4 \cdot 15 = 3(x + 10)$$
   $$60 = 3x + 30$$
   $$-30 = -30$$
   $$30 = 3x$$
   $$\frac{30}{3} = \frac{3x}{3}$$
   $$10 = x$$

Another way to solve this is to make the two fractions have a common denominator of 15 and then use the fact that the numerators will then be equal.

$$\frac{4}{3} = \frac{x + 10}{15}$$

$$\frac{20}{15} = \frac{x + 10}{15}$$

$$20 = x + 10$$

$$-10 = -10$$

$$10 = x$$

The correct choice is **(3)**.

5. Graph both functions on your graphing calculator.

On the TI-84:

On the TI-Nspire:

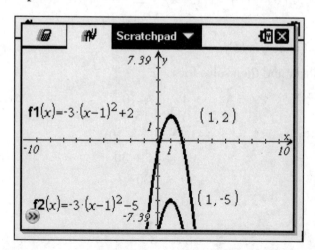

In a parabola that opens downward, the vertex is the maximum point. The vertex of the graph of function f is $(1, 2)$. The vertex of the graph of function g is $(1, -5)$. So, the vertex of the graph of $g(x)$ is 7 units below the vertex of function $f(x)$.

This can also be done without a graphing calculator by using the properties of the vertex form of an equation for a parabola.

When a function is in the form $f(x) = a(x - h)^2 + k$, the graph will be a parabola with its vertex at (h, k). Function f is in exactly this form, so its vertex is $(1, 2)$. Function g can be expressed as $g(x) = -3(x - 1)^2 + (-5)$, so its vertex is $(1, -5)$. The point $(1, -5)$ is 7 units below the point $(1, 2)$.

The correct choice is (**1**).

6. The percent of boys who chose Niagara Falls can be calculated by dividing the number of boys who chose Niagara Falls by the total number of boys. First find the total number of boys.

$$56 + 74 + 103 = 233$$

Now find the percent of boys.

$$\frac{56}{233} \approx 0.24 = 24\%$$

The correct choice is (**2**).

7. Compare the shape of this graph to the shape of typical graphs for each of the answer choices.

Choice (1):
The graph of a linear function is a straight line.

 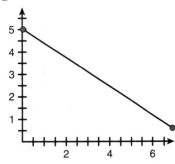

Choice (2):

The graph of an exponential function looks like a playground slide.

Choice (3):

The graph of a square root function looks like half of a sideways parabola.

Choice (4):

The graph of an absolute value function looks like the letter "V."

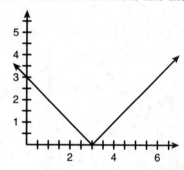

The given graph most resembles a portion of the graph of an exponential function.

The correct choice is **(2)**.

8. The difference of perfect squares factoring pattern is $a^2 - b^2 = (a - b)(a + b)$. Because both terms of the given expression are perfect squares, the expression can be written as the difference of perfect squares.

$$16x^2 - 81 = (4x)^2 - 9^2$$
$$= (4x - 9)(4x + 9)$$

The correct choice is (3).

9. Depending on what the owner of the business is looking to measure, any of these choices could be appropriate. Since in this problem the owner wants to know the average amount of time his team spends mowing one lawn, he would calculate the answer by dividing the total number of hours spent by the total number of lawns mowed. For example, if the workers took 8 hours to mow 4 lawns, the average amount of time to mow one lawn would be

$$\frac{8 \text{ hours}}{4 \text{ lawns}} = \frac{8}{4} = 2 \text{ hours per lawn.}$$

The correct choice is (4).

10. When the ball hits the ground, the height of the ball above the ground will be 0. To find when this happens, solve the equation $0 = -16t^2 + 64t + 80$.

$$0 = -16t^2 + 64t + 80$$
$$0 = -16\left(t^2 - 4t - 5\right)$$
$$0 = -16(t - 5)(t + 1)$$
$$t - 5 = 0 \qquad t + 1 = 0$$
$$t = 5 \qquad \text{or } t = -1$$

Since time must be positive, the solution is $t = 5$.

This answer can also be determined by graphing the function on a graphing calculator and finding the x-intercept.

On the TI-84:

On the TI-Nspire:

The correct choice is (**1**).

11. The given equation can be expressed in a form like the answer choices by using the completing the square process.

The first step is to eliminate the coefficient from the right-hand side by adding 18 to both sides of the equation.

$$y = x^2 + 24x - 18$$
$$+18 = +18$$
$$y + 18 = x^2 + 24x$$

Next, square half the coefficient of the x-term and add that to both sides of the equation.

$$\left(\frac{24}{2}\right)^2 = 12^2 = 144$$

$$y + 18 + 144 = x^2 + 24x + 144$$
$$y + 162 = x^2 + 24x + 144$$

The right-hand side of the equation is now a perfect square trinomial and can be factored.

$$y + 162 = x^2 + 24x + 144$$
$$y + 162 = (x + 12)^2$$

Isolate the y by subtracting 162 from both sides of the equation.

$$y + 162 = (x + 12)^2$$
$$-162 = -162$$
$$y = (x + 12)^2 - 162$$

This question can also be solved without completing the square by just simplifying each of the four answer choices to see which is equivalent to $y = x^2 + 24x - 18$.

Testing choice (1):

$y = (x + 12)^2 - 162$

$y = x^2 + 24x + 144 - 162$

$y \overset{\checkmark}{=} x^2 + 24x - 18$

Testing choice (2):

$y = (x + 12)^2 + 126$

$y = x^2 + 24x + 144 + 126$

$y = x^2 + 24x + 270$

$y \neq x^2 + 24x - 18$

Testing choice (3):

$y = (x - 12)^2 - 162$

$y = x^2 - 24x + 144 - 162$

$y = x^2 - 24x - 18$

$y \neq x^2 + 24x - 18$

Testing choice (4):

$y = (x - 12)^2 + 126$

$y = x^2 - 24x + 144 + 126$

$y = x^2 - 24x + 270$

$y \neq x^2 + 24x - 18$

The correct choice is (**1**).

12. First multiply the factors together.

$$(x)(x - 5)(2x + 3) = (x^2 - 5x)(2x + 3)$$
$$= 2x^3 + 3x^2 - 10x^2 - 15x$$
$$= 2x^3 - 7x^2 - 15x$$

This polynomial has three terms: $2x^3$, $7x^2$, and $15x$.

The constant term is 0 since there is no term that is not multiplied by x or a power of x. Because the highest exponent is a 3, the degree of this polynomial is 3.

The leading coefficient is the number that is multiplied by the highest power of x. Since the term $2x^3$ has the highest power of x, the leading coefficient is 2.

The correct choice is (**2**).

13. One of the laws of exponents is $x^{ab} = \left(x^a\right)^b$. Apply this to the given expression.

$$P(t) = 3810(1.0005)^{7t}$$

$$= 3810\left(1.0005^7\right)^t$$

$$= 3810(1.0035)^t$$

Another way to solve this problem is to pick any value you like for t, such as $t = 10$. Then calculate $P(10)$ in the original function.

$$P(10) = 3810(1.0005)^{7 \cdot 10}$$

$$= 3810(1.0005)^{70}$$

$$= 3810(1.03561065)$$

$$\approx 3945.7$$

Then test for $t = 10$ in each of the answer choices. When you test choice (2), you find approximately the same result.

$$P(10) = 3810(1.0035)^{10}$$

$$= 3810(1.035556427)$$

$$\approx 3945.5$$

The correct choice is (**2**).

14. The x-coordinates of the intersection points of the line and the curve are the solutions to $f(x) = g(x)$. Since the line intersects the curve at $(-2, 5)$, $(-1, 4)$, and $(2, 1)$, the three x-values that make $f(x) = g(x)$ are -2, -1, and 2. Of the four choices, each of these three values appear. However, 3 is not one of the solutions.

The correct choice is (**3**).

15. In a box plot, the left endpoint represents the minimum value. For this problem, the minimum value is 1. The right endpoint represents the maximum value. For this problem, the maximum value is 8. The range is the difference between the maximum value and the minimum value, which is $8 - 1 = 7$.

The correct choice is (**1**).

16. Factor the given expression by first factoring out the greatest common factor of 2.

$$2x^2 + 10x + 12 = 2\left(x^2 + 5x + 6\right)$$
$$= 2(x + 3)(x + 2)$$

Since this is choice (4) and the question is asking which choice is *not* equivalent, choice (4) can be eliminated.

To find the other two equivalent expressions, distribute the 2 to each of the factors.

$$2(x + 3)(x + 2) = (2x + 6)(x + 2)$$

So, choice (2) can be eliminated.

$$2(x + 3)(x + 2) = (x + 3)2(x + 2)$$
$$= (x + 3)(2x + 4)$$

So, choice (1) can be eliminated.

Another way to solve this problem is to multiply out each of the four choices to see which one is not equivalent to the given expression.

Testing choice (1):

$$(2x + 4)(x + 3) = 2x^2 + 6x + 4x + 12$$
$$= 2x^2 + 10x + 12$$

Testing choice (2):

$$(2x + 6)(x + 2) = 2x^2 + 4x + 6x + 12$$
$$= 2x^2 + 10x + 12$$

Testing choice (3):

$$(2x + 3)(x + 4) = 2x^2 + 8x + 3x + 12$$
$$= 2x^2 + 11x + 12$$
$$\neq 2x^2 + 10x + 12$$

Testing choice (4):

$$2(x + 3)(x + 2) = (2x + 6)(x + 2)$$
$$= 2x^2 + 4x + 6x + 12$$
$$= 2x^2 + 10x + 12$$

The correct choice is **(3)**.

17. Based on the data in the table, $r(-2) = -16$ is the minimum of function r. For function q, you can graph the function on a graphing calculator and find the coordinates of the minimum point using the minimum function.

On the TI-84:

On the TI-Nspire:

The minimum point is located at $(-1, -9)$, so the minimum value for function q is -9.

Note that the x-coordinate is not the minimum; the y-coordinate is the minimum.

Even without using a graphing calculator, you can calculate the x-coordinate of the minimum of function q with the equation $x = -\dfrac{b}{2a}$.

$$x = -\frac{b}{2a}$$
$$= -\frac{2}{2 \cdot 1}$$
$$= -1$$

Then calculate the value of q when $x = 1$.

$$q(-1) = (-1)^2 + 2(-1) - 8$$
$$= 1 - 2 - 8$$
$$= -9$$

Since $-16 < -9$, function r has the smallest minimum value.

The correct choice is (**3**).

18. At $t = 0$, the child is 10 yards away from home. As the child gets closer to home, the distance from the child to home gets smaller. So, an interval in which the child is constantly moving closer to home will be always decreasing. This happens in the interval between $t = 0$ and $t = 2$. The other intervals have some parts where the distance is either increasing or staying the same.

The correct choice is (**1**).

19. Substitute $n = 2$ into the second part of the recursive definition.

$$a_2 = 3 + 2(a_{2-1})^2$$
$$= 3 + 2(a_1)^2$$
$$= 3 + 2 \cdot 6^2$$
$$= 3 + 2 \cdot 36$$
$$= 3 + 72$$
$$= 75$$

The correct choice is (**1**).

20. If the width is w feet and the length is 7 feet more than the width of a rectangular patio, the length can be expressed as $w + 7$ feet. The area of a rectangle is length times width.

$$A(w) = lw$$
$$= (w + 7)w$$
$$= w^2 + 7w$$

The correct choice is (**2**).

21. A sketch of the graph could look like the following.

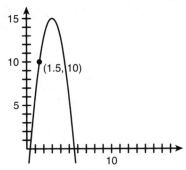

The domain is the set of x-coordinates of all the points on the graph. This domain is not limited to integers since there are points that have x-coordinates that are not integers, such as (1.5, 10). Numbers between integers qualify as real numbers. Since the graph starts at $x = 0$, there are also no negative x-values. Negative x-values would be used for moments before the dolphin begins jumping. So, positive real numbers would be most appropriate for the domain of this graph.

If the question instead asked for the range of the graph, then real numbers, including negative values, would be appropriate since there are times where the dolphin is under water. The y-coordinates of the points representing those underwater moments would be negative.

The correct choice is (**4**).

22. Solve the given system of equations with the elimination method. First multiply both sides of the second equation by -1 and then add the first equation to that result.

$$x - 4y = -10$$
$$-1(x + y) = -1(5)$$

$$\begin{aligned} x - 4y &= -10 \\ + \quad -x - y &= -5 \\ \hline \frac{-5y}{-5} &= \frac{-15}{-5} \\ y &= 3 \end{aligned}$$

Substitute $y = 3$ into either of the two original equations to solve for x.

$$x + 3 = 5$$
$$-3 = -3$$
$$x = 2$$

The solution to the given system is $(2, 3)$.

The answer choices are much simpler systems since in each of them, one of the equations has just one variable. Also notice that the other equation in each choice matches one of the equations in the given system. So, all that needs to be done is to check each choice to see if the solution to the one-variable equation matches either $x = 2$ or $y = 3$.

Checking choice (1):

$$\frac{5x}{5} = \frac{10}{5}$$
$$x = 2$$

Checking choice (2):

$$\frac{-5y}{-5} = \frac{-5}{-5}$$
$$y = 1$$

Checking choice (3):

$$\frac{-3x}{-3} = \frac{-30}{-3}$$
$$x = 10$$

Checking choice (4):

$$\frac{-5y}{-5} = \frac{-5}{-5}$$
$$y = 1$$

The correct choice is **(1)**.

23. The range is the set of y-coordinates of all the points on the graph of a function. Graph this function on a graphing calculator.

On the TI-84:

On the TI-Nspire:

The lowest y-coordinate of the graph is -4.5. There is no highest y-coordinate since both branches of the parabola go up forever.

One way to express this range is $y \geq -4.5$. The answer choices, however, use interval notation.

The only difference between choices (2) and (4) is that choice (2) has a parenthesis "(" on the left side of the interval and choice (4) has a square bracket "[."

The square bracket "[" is used when the interval does contain the first number. In this case, since the minimum point of the parabola is -4.5, the square bracket "[" is the appropriate notation.

The correct choice is **(4)**.

24. A geometric sequence is one in which each term is equal to the previous term multiplied by some constant called the common ratio. For example, 3, 6, 12, 24, . . . is a geometric series with a common ratio of 2.

   If the common ratio is called r and the first term of the sequence is 5, then the second term will equal $5r$ and the third term will equal $5r \cdot r = 5r^2$.

   One way to solve this problem is to find the value of r that makes $5r^2$ equal to 245.

$$\frac{5r^2}{5} = \frac{245}{5}$$
$$r^2 = 49$$
$$r = \pm 7$$

   There are two valid answers, but only one of them is among the four answer choices.

   The correct choice is **(1)**.

PART II

25. To determine $g(-2)$, substitute -2 for x in the given equation.

$$\begin{aligned} g(-2) &= -4(-2)^2 - 3(-2) + 2 \\ &= -4 \cdot 4 + 6 + 2 \\ &= -16 + 6 + 2 \\ &= -8 \end{aligned}$$

26. The commutative property of arithmetic says that $a + b = b + a$. This can be applied to an expression involving negatives, like $-5a + 7 = 7 + (-5a) = 7 - 5a$.

The first step the student used was to rearrange the second and third terms of the right-hand side of the equation. So, the student used the commutative property of addition.

This is correct because using the commutative property of addition does not change the value of the right-hand side of the equation.

Most students would skip this as their first step and instead combine like terms to get $3a + 6 = 9 - 5a$. However, this commutative property step actually allows students to combine like terms.

27. Since there are no exponents greater than or equal to 2, this is a linear equation and the graph will be a line. Only two points are needed to graph a line. To be sure, though, make a chart with more values.

Pick any value you want for x, and substitute into the equation to solve for the corresponding y value. For example, choose $x = 2$.

$$\begin{aligned} 2y &= -3(2) - 2 \\ 2y &= -6 - 2 \\ \frac{2y}{2} &= \frac{-8}{2} \\ y &= -4 \end{aligned}$$

x	y
0	-1
1	-2.5
2	-4
3	-5.5
4	-7

Plot the points on the graph, and connect them to draw the entire line.

 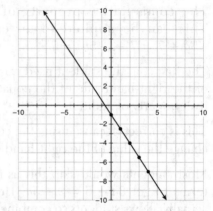

This could also have been done by first converting to slope-intercept form by dividing both sides of the equation by 2.

$$\frac{2y}{2} = \frac{-3x - 2}{2}$$

$$y = -\frac{3}{2}x - 1$$

The -1 is the y-intercept and the $-\frac{3}{2}$ is the slope. Graph the point $(0, -1)$. Then, using the slope of $-\frac{3}{2}$, move 2 units right and 3 units down to get the point $(2, -4)$. Join these two points to form the line.

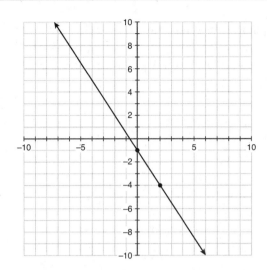

Since $(2, -4)$ is a point on the line, -4 is the value of k.

28. First multiply both sides of the equation by t.

$$at = v_f - v_i$$

Next, add v_i to both sides of the equation to isolate the v_f term.

$$at = v_f - v_i$$
$$+v_i = +v_i$$
$$at + v_i = v_f$$

29. Solve the inequality the same way you would solve it if there were an "=" sign instead of a "<." However, be sure to reverse the direction of the inequality sign when you multiply or divide both sides of the inequality by a negative number.

$$\frac{3}{5}x + \frac{1}{3} < \frac{4}{5}x - \frac{1}{3}$$

$$-\frac{1}{3} = -\frac{1}{3}$$

$$\frac{3}{5}x < \frac{4}{5}x - \frac{2}{3}$$

$$-\frac{4}{5}x = -\frac{4}{5}x$$

$$-\frac{1}{5}x < -\frac{2}{3}$$

$$-5\left(-\frac{1}{5}x\right) > -5\left(-\frac{2}{3}\right)$$

$$x > \frac{10}{3}$$

30. No. The product of two irrational numbers can be rational. For example, $\sqrt{2} \cdot \sqrt{2} = 2$. The number $\sqrt{2}$ is irrational, and the number 2 is rational.

31. Rearrange the terms to solve for x.

$$6x^2 - 42 = 0$$

$$+42 = +42$$

$$\frac{6x^2}{6} = \frac{42}{6}$$

$$x^2 = 7$$

$$\sqrt{x^2} = \pm\sqrt{7}$$

$$x = \pm\sqrt{7}$$

32. Make a table of values for the piecewise functions. For negative values, use the top equation. For values between 0 and 5, use the bottom equation.

x	$h(x)$
-2	$2(-2) - 3 = -7$
-1	$2(-1) - 3 = -5$
0	$0^2 - 4(0) - 5 = -5$
1	$1^2 - 4(1) - 5 = -8$
2	$2^2 - 4(2) - 5 = -9$
3	$3^2 - 4(3) - 5 = -8$
4	$4^2 - 4(4) - 5 = -5$
5	$5^2 - 4(5) - 5 = 0$

Plot these eight points with closed circles.

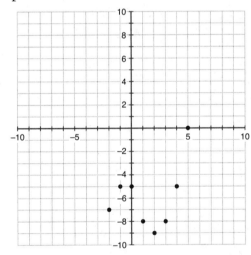

Before drawing the curves, evaluate the top equation at 0 to get $2(0) - 3 = -3$. Even though the point $(0, -3)$ is not part of the graph since 0 is part of the other "piece," this point is the boundary between the two pieces. Plot $(0, -3)$ with an open circle to represent this.

Now draw the curves.

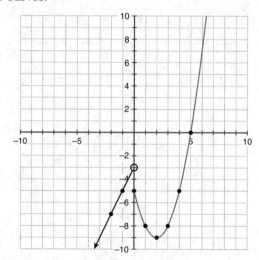

The TI-Nspire has the built-in capability for graphing piecewise functions, though it does not clearly show the open circle boundary point.

PART III

33. Rewrite both inequalities in slope-intercept form.

$$y \geq -2x + 8$$
$$y < 3x + 5$$

Graph the line $y = -2x + 8$ with a solid line because it contains the sign "$\geq$" and not "$>$." Test the ordered pair $(0, 0)$ to determine if it makes the inequality $y \geq -2x + 8$ true.

$$0 \geq -2(0) + 8$$
$$0 \geq 0 + 8$$
$$0 \geq 8$$

Since $(0, 0)$ does not make the first inequality true, shade the side of the line $y = -2x + 8$ that does not contain $(0, 0)$.

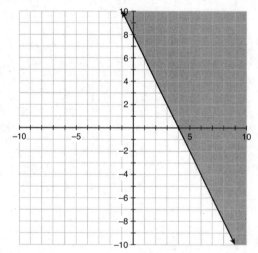

Graph the line $y = 3x + 5$ with a dotted line because it contains the sign "$<$" and not "$\leq$." Test the ordered pair $(0, 0)$ to determine if it makes the inequality $y = 3x + 5$ true.

$$0 < 3(0) + 5$$
$$0 < 0 + 5$$
$$0 < 5$$

Since (0, 0) does make the second inequality true, shade (with a different style of shading) the side of the line $y = 3x + 5$ that does contain (0, 0).

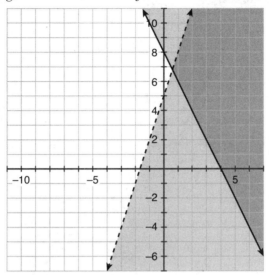

The entire double-shaded region is the solution set to the system of inequalities.

Systems of inequalities can also be graphed on a graphing calculator.

On the TI-84:

On the TI-Nspire:

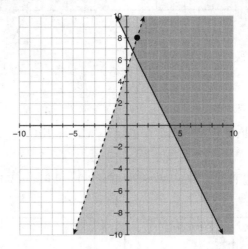

Since the point $(1, 8)$ is on the dotted line, it is not part of the solution set.

34. The formula for exponential growth for a given growth rate is $A(t) = P(1 + r)^t$, where P is the original investment and r is the growth rate. For this problem, the function can be written as follows:

$$A(t) = 5000(1 + 0.012)^t = 5000(1.012)^t$$

Subtract the values when $t = 32$ and $t = 17$ to find the solution.

$$A(32) - A(17) = 5000(1.012)^{32} - 5000(1.012)^{17}$$
$$\approx 7324 - 6124$$
$$= 1200$$

To the *nearest dollar*, the investment will be worth $1200 more when Alexander turns 32 than when he turns 17.

35. A linear regression equation, or line of best fit, for a data set can be calculated with a graphing calculator.

For the TI-84:

To turn diagnostics on, press [2nd] [0], scroll down to select DiagnosticOn, and press [ENTER].

```
CATALOG             ▯
 DependAuto
 det(
 DiagnosticOff
▸DiagnosticOn
 dim(
 Disp
 DispGraph
```

Press [STAT] [1]. Enter the x-values into L1 and the y-values into L2.

```
L1      L2      L3      3

0       293     ▬▬▬▬▬
0       263
1       244
1       224
3       185
4       170
7       219

L3(1)=
```

Press [STAT] [right arrow] [4] to get the LinReg function. Select Calculate.

```
        LinReg
y=ax+b
a=-7.756345178
b=246.3380711
r²=.7685525674
r=-.8766712995
```

For the TI-Nspire:
From the home screen, select the Add Lists & Spreadsheet icon. In the first row, label the A column "hours" and the B column "score." Enter the hours in cells A1 to A10. Enter the scores in cells B1 to B10.

	A hours	B score	C	D
◆				
1	0	293		
2	0	263		
3	1	244		
4	1	224		
5	3	185		

C1

Move the cursor to cell C1. Press [menu] [4] [1] [3]. Set X List to "hours" and Y List to "score." Select OK.

	B score	C	D	E =LinRegM
◆				
1	293		Title	Linear Re..
2	263		RegEqn	m*x+b
3	244		m	-7.75635
4	224		b	246.338
5	185		r²	0.768553

E1 ="Linear Regression (mx+b)"

The line of best fit is $y = -7.76x + 246.34$.

The correlation coefficient (r) is a measure of how close the points are to the line of best fit. If the line of best fit has a positive slope, the correlation coefficient will be positive. If the line of best fit has a negative slope, the correlation coefficient will be negative. The closer the correlation coefficient is to 1 (or to -1), the closer to the line of best fit the points will lie.

For this data set, the correlation coefficient is -0.88. Because the correlation coefficient is negative, the line of best fit must have a negative slope.

36. Make a table of values for times 0 to 17, following the increases described in the problem.

Elapsed Time	Accumulated Snowfall
0	0
1	0.5
2	1
3	1.5
4	2
5	3
6	4
7	5
8	6
9	7
10	8
11	8
12	8
13	8
14	8.5
15	9
16	9.5
17	10

Plot those seventeen points, and connect the points to form the graph.

The average rate of snowfall is the total amount of snowfall divided by the total number of hours. For this problem, that is $\frac{10}{17} \approx 0.59$ inches per hour.

PART IV

37. There are 12 chickens, so the first equation is $A + D = 12$. The chickens cost 35 dollars, so the other equation is $3.75A + 2.50D = 35$.

Write the system of equations.

$$A + D = 12$$
$$3.75A + 2.50D = 35$$

This system can be solved with either the elimination method or with the substitution method.

Elimination method:
Multiply both sides of the top equation by -2.50, and add the equations together.

$$-2.50A - 2.50D = -30$$
$$+ \quad 3.75A + 2.50D = 35$$
$$\frac{1.25A}{1.25} = \frac{5}{1.25}$$
$$A = 4$$

Solve for D by plugging $A = 4$ into one of the original equations.

$$4 + D = 12$$
$$D = 8$$

Substitution method:
To use this method, one of the equations must be changed so that one of the variables is isolated. This can be done most easily with the first equation. Subtract D from both sides to make the equation $A = 12 - D$. Next, substitute into the second equation and solve for D.

$$3.75(12 - D) + 2.50D = 35$$
$$45 - 3.75D + 2.50D = 35$$
$$45 - 1.25D = 35$$
$$-45 = -45$$
$$\frac{-1.25D}{-1.25} = \frac{-10}{-1.25}$$
$$D = 8$$

Then solve for A by substituting $D = 8$ into the first equation.

$$A + 8 = 12$$
$$A = 4$$

Allysa bought 4 Americana chickens and 8 Delaware chickens.

With 4 Americana chickens and 8 Delaware chickens, Allysa will get $4 \cdot 2 + 8 \cdot 1 = 16$ eggs per day. After one week, she will have $16 \cdot 7 = 112$ eggs. Divide this by 12 to see how many dozens of eggs she will have in one week.

$$\frac{112}{12} \approx 9.3$$

Since she sells eggs only by the full dozen, Allysa can sell 9 dozen for a total of $9 \cdot \$2.50 = \$22.50$.

Topic	Question Numbers	Number of Points	Your Points	Your Percentage
1. Polynomials	3, 8, 12, 16	$2 + 2 + 2 + 2 = 8$		
2. Properties of Algebra	4, 26, 28	$2 + 2 + 2 = 6$		
3. Functions	2, 14, 21, 23, 25	$2 + 2 + 2 + 2 + 2 = 10$		
4. Creating and Interpreting Equations	1, 20	$2 + 2 = 4$		
5. Inequalities	29, 33	$2 + 4 = 6$		
6. Sequences and Series	19, 24	$2 + 2 = 4$		
7. Systems of Equations	22, 37	$2 + 6 = 8$		
8. Quadratic Equations and Factoring	5, 10, 11, 17, 31	$2 + 2 + 2 + 2 + 2 = 10$		
9. Regression	35	4		
10. Exponential Equations	7, 13, 34	$2 + 2 + 4 = 8$		
11. Graphing	18, 27, 32, 36	$2 + 2 + 2 + 4 = 10$		
12. Statistics	6, 15	$2 + 2 = 4$		
13. Number Properties	30	2		
14. Unit Conversions	9	2		

HOW TO CONVERT YOUR RAW SCORE TO YOUR ALGEBRA I REGENTS EXAMINATION SCORE

The accompanying conversion chart must be used to determine your final score on the August 2019 Regents Examination in Algebra I. To find your final exam score, locate in the column labeled "Raw Score" the total number of points you scored out of a possible 86 points. Since partial credit is allowed in Parts II, III, and IV of the test, you may need to approximate the credit you would receive for a solution that is not completely correct. Then locate in the adjacent column to the right the scale score that corresponds to your raw score. The scale score is your final Algebra I Regents Examination score.

Regents Examination in Algebra I—August 2019
Chart for Converting Total Test Raw Scores to Final
Exam Scores (Scale Scores)

Raw Score	Scale Score	Performance Level	Raw Score	Scale Score	Performance Level	Raw Score	Scale Score	Performance Level
86	100	5	57	81	4	28	66	3
85	99	5	56	81	4	27	65	3
84	98	5	55	81	4	26	64	2
83	96	5	54	81	4	25	63	2
82	95	5	53	80	4	24	61	2
81	94	5	52	80	4	23	60	2
80	93	5	51	80	4	22	58	2
79	92	5	50	80	4	21	57	2
78	91	5	49	79	3	20	55	2
77	90	5	48	79	3	19	53	1
76	90	5	47	79	3	18	51	1
75	89	5	46	78	3	17	49	1
74	88	5	45	78	3	16	47	1
73	87	5	44	78	3	15	45	1
72	87	5	43	77	3	14	43	1
71	86	5	42	77	3	13	41	1
70	86	5	41	76	3	12	38	1
69	86	5	40	76	3	11	36	1
68	85	5	39	75	3	10	33	1
67	84	4	38	75	3	9	30	1
66	84	4	37	74	3	8	27	1
65	84	4	36	74	3	7	24	1
64	83	4	35	73	3	6	21	1
63	83	4	34	72	3	5	18	1
62	83	4	33	71	3	4	15	1
61	82	4	32	70	3	3	11	1
60	82	4	31	70	3	2	8	1
59	82	4	30	69	3	1	4	1
58	82	4	29	68	3	0	0	1

NOTES

NOTES

NOTES

NOTES